The PENROSE TRANSFORM

The PENROSE TRANSFORM

Its Interaction with Representation Theory

Robert J. Baston
Michael G. Eastwood

Dover Publications, Inc., Mineola, New York

Bibliographical Note

This Dover edition, first published in 2016, is a slightly corrected and unabridged republication of the work originally published in the Oxford Mathematical Monographs Series by The Clarendon Press, Oxford, in 1989. A new Preface written by the authors has been specially prepared for this Dover edition.

Library of Congress Cataloging-in-Publication Data

Baston, Robert J.
  The Penrose transform : its interaction with representation theory / Robert J. Baston and Michael G. Eastwood. — Dover edition.
    pages cm
  Slightly corrected and unabridged.
  Originally published: Oxford : Clarendon Press, 1989.
  Includes bibliographical references and index.
  ISBN-13: 978-0-486-79729-8 — ISBN-10: 0-486-79729-5
  1. Penrose transform. 2. Mathematical physics. 3. Representations of groups.
4. Geometry, Differential. I. Eastwood, Michael G. II. Title.

QC20.7.D52B37 2015
530.1'5636—dc23

2015008915

Manufactured in the United States
79729501   2016
www.doverpublications.com

To Wendy and Meryl

PREFACE TO THE DOVER EDITION

It is now 27 years since the original edition of this book and a great deal has happened since then. Indeed, it is impossible to summarise the full extent of the developments in a short preface. For that, at least another volume would be required. Instead, we shall indicate just two areas in which there has been substantial progress.

BGG Machinery in Parabolic Geometry. Bernstein–Gelfand–Gelfand complexes were introduced into differential geometry in [5], although in this article they were created *ad hoc* and the connection to the purely algebraic BGG complexes of generalised Verma modules was realised only later. Looking back in [5], one now sees the BGG differential complexes on $\mathbb{CP}_3$ on page 351 and the BGG complexes on $\mathrm{Gr}_2(\mathbb{C}^4)$ on page 352.

By the time this book was written, the geometric interpretation of the BGG complexes on complex flag manifolds G/P for G semisimple and P a parabolic subgroup was well understood. This is the topic of Chapter 8. It is already clear that there are real versions of these differential complexes, whenever this makes sense. For example, using the same notation as in Chapter 8, there are differential complexes

$$\overset{a\ \ \ b\ \ \ c}{\times\!\!-\!\!\bullet\!\!-\!\!\bullet} \to \overset{-a-2\ \ \ a+b+1\ \ \ c}{\times\!\!-\!\!\!-\!\!\!-\!\!\bullet\!\!-\!\!\bullet} \to \overset{-a-b-3\ \ \ a\ \ \ b+c+1}{\times\!\!-\!\!\!-\!\!\!-\!\!\bullet\!\!-\!\!\bullet} \to \overset{-a-b-c-4\ \ \ a\ \ \ b}{\times\!\!-\!\!\!-\!\!\!-\!\!\bullet\!\!-\!\!\bullet} \to 0$$

on $\mathbb{RP}_3$ as a homogeneous space for $\mathrm{SL}(4,\mathbb{R})$. When $a = b = c = 0$, it is the de Rham complex and a resolution of $\mathbb{R}$, the locally constant functions. More generally, for any non-negative integers a, b, c, this complex is locally exact and a resolution of the locally constant sections of the homogeneous bundle $\overset{a\ \ \ b\ \ \ c}{\bullet\!\!-\!\!\bullet\!\!-\!\!\bullet}$ on $\mathbb{RP}_3$ obtained by inducing up to $\mathbb{RP}_3 = \mathrm{SL}(4,\mathbb{R})/P$, the real representation $\overset{a\ \ \ b\ \ \ c}{\bullet\!\!-\!\!\bullet\!\!-\!\!\bullet}$ of $\mathrm{SL}(4,\mathbb{R})$ restricted to P. Here, 'locally constant' means locally *covariantly* constant with respect to an invariantly defined connection on the bundle $\overset{a\ \ \ b\ \ \ c}{\bullet\!\!-\!\!\bullet\!\!-\!\!\bullet}$ and, more generally, if E is any finite-dimensional real representation of a real Lie group G, then the induced vector bundle

$$\mathbb{E} = G \times_P E \text{ on } G/P$$

acquires a flat G-invariant connection by dint of the trivialisation

$$G \times_P E \cong G/P \times E$$
$$[g, \alpha] \longmapsto (g \bmod P, g\alpha).$$

Parabolic differential geometry is a relatively new field [2] developed over the past 25 years in which geometries are locally modelled on real homogeneous spaces G/P for G semisimple and P parabolic. Several classical differential geometries are included in the parabolic realm and, for example, $\mathbb{RP}_3 = \mathrm{SL}(4, \mathbb{R})/P$ is the 'flat model' of 3-dimensional projective differential geometry. More explicitly, one says that two torsion-free affine connections ∇_a and $\hat{\nabla}_a$ on a smooth manifold M are *projectively equivalent* if and only if they have the same local geodesics as unparameterised curves. Operationally, it means that

$$\hat{\nabla}_a \phi_b = \nabla_a \phi_b - \Upsilon_a \phi_b - \Upsilon_b \phi_a \quad \text{for some 1-form } \Upsilon_a.$$

The connections in question need not arise from any metric but there are non-trivial examples of projective equivalence even when they do. For example [9], the metrics

$$dx^2 + dy^2 \quad \text{and} \quad \frac{(1+y^2)\, dx^2 - 2xy\, dx\, dy + (1+x^2)\, dy^2}{(1+x^2+y^2)^2}$$

are projectively equivalently. Indeed, the second of these is the round hemisphere and extends to the round metric on $\mathbb{RP}_2$. This explains why the round $\mathbb{RP}_2$ is just as good, if not better, than flat $\mathbb{R}^2$ as the 'flat model' of 2-dimensional projective differential geometry. Similarly, the projective structure induced by the round metric on $\mathbb{RP}_3$ is, in fact, the best 'flat model' in dimension 3 (and its projective symmetries are induced by the natural action of $\mathrm{SL}(4, \mathbb{R})$).

It turns out that many of the features of G/P persist in the curved setting. For example, the operators in the BGG complex are replaced by curved versions thereof [3,4]—there are invariant operators between appropriate bundles with the same symbol as in the flat model. The meaning of invariance has to be upgraded—roughly speaking, it means that the operators in question see only the local geometric structure specified by the parabolic theory. In projective differential geometry it means that, although one might write the operators in terms of an arbitrarily chosen connection in the projective class, these operators are, in fact, independent of this choice. These curved *BGG sequences* often have geometric significance. For example, the curved sequence

$$\overset{0\ \ 1\ \ 0}{\times\!\!-\!\!\bullet\!\!-\!\!\bullet} \overset{\nabla}{\longrightarrow} \overset{-2\ \ 2\ \ 0}{\times\!\!-\!\!\bullet\!\!-\!\!\bullet} \overset{\nabla^2}{\longrightarrow} \overset{-4\ \ 0\ \ 2}{\times\!\!-\!\!\bullet\!\!-\!\!\bullet} \overset{\nabla}{\longrightarrow} \overset{-5\ \ 0\ \ 1}{\times\!\!-\!\!\bullet\!\!-\!\!\bullet}$$

on a 3-dimensional projective differential geometry that just happens to contain the Levi-Civita connection a Riemannian metric g_{ab}, controls the deformation theory of that metric. Specifically, the first operator can be identified as the Killing operator $X \mapsto \mathcal{L}_X g_{ab}$, the second as the linearisation around g_{ab} of the Ricci curvature of the perturbed metric $g_{ab} + \epsilon h_{ab}$, and the third as $R_{ab} \mapsto 2\nabla^a R_{ab} - \nabla_b R^a{}_a$, whose kernel is the symmetric tensors satisfying the Bianchi identity. (It is interesting to contemplate the corresponding sequences in other dimensions,

$$\overset{0\ \ 1}{\times\!\!-\!\!\bullet} \overset{\nabla}{\longrightarrow} \overset{-2\ \ 2}{\times\!\!-\!\!\bullet} \overset{\nabla^2}{\longrightarrow} \overset{-4\ \ 0}{\times\!\!-\!\!\bullet} \qquad \text{in dimension 2}$$

and

$$\overset{0\ \ 1\ \ 0\ \ 0}{\times\!\!-\!\!\bullet\!\!-\!\!\bullet\!\!-\!\!\bullet}\cdots\overset{0}{-\!\!\bullet} \overset{\nabla}{\longrightarrow} \overset{-2\ \ 2\ \ 0\ \ 0}{\times\!\!-\!\!\bullet\!\!-\!\!\bullet\!\!-\!\!\bullet}\cdots\overset{0}{-\!\!\bullet} \overset{\nabla^2}{\longrightarrow} \overset{-4\ \ 0\ \ 2\ \ 0}{\times\!\!-\!\!\bullet\!\!-\!\!\bullet\!\!-\!\!\bullet}\cdots\overset{0}{-\!\!\bullet} \overset{\nabla}{\longrightarrow}\ \cdots$$

in dimension ≥ 4, correctly realising the Riemannian curvature as the Gaußian scalar in 2 dimensions and enjoying the irreducible symmetries

$$R_{abcd} = R_{[ab][cd]} \qquad R_{[abc]d} = 0$$

in dimension ≥ 4.) Concerning deformations of projective structures themselves, one should firstly contemplate the BGG sequence

$$\overset{1\ \ 0\ \ 0\ \ 0}{\times\!\!-\!\!\bullet\!\!-\!\!\bullet\!\!-\!\!\bullet}\cdots\overset{1}{-\!\!\bullet} \overset{\nabla^2}{\longrightarrow} \overset{-3\ \ 2\ \ 0\ \ 0}{\times\!\!-\!\!\bullet\!\!-\!\!\bullet\!\!-\!\!\bullet}\cdots\overset{1}{-\!\!\bullet} \overset{\nabla}{\longrightarrow} \overset{-4\ \ 1\ \ 1\ \ 0}{\times\!\!-\!\!\bullet\!\!-\!\!\bullet\!\!-\!\!\bullet}\cdots\overset{1}{-\!\!\bullet} \overset{\nabla}{\longrightarrow}\ \cdots,$$

though the first BGG operator $X^a \mapsto \left(\nabla_{(a}\nabla_{b)}X^c + \frac{1}{n-1}R_{ab}X^c\right)_\circ$, where '$\circ$' means to take the trace-free part, has to be modified

$$X^a \mapsto \left(\nabla_{(a}\nabla_{b)}X^c + \frac{1}{n-1}R_{ab}X^c\right)_\circ + W_{d(a}{}^c{}_{d)}X^d,$$

by the projective Weyl curvature $W_{ab}{}^c{}_d \equiv (R_{ab}{}^c{}_d)_\circ$, in order to obtain the projective Killing operator (with kernel the infinitesimal projective symmetries). Nevertheless, one sees the BGG sequences and machinery at the heart of projective differential geometry.

These remarks hold true throughout the parabolic realm (a glimpse of which can be found in [2, Chapter 4], 'A panorama of examples'). Conformal differential geometry is another classical geometry within this realm, with ●—✕—● $= \mathrm{Gr}_2(\mathbb{R}^4)$ as the flat model of signature $(2,2)$ conformal spin geometry in 4 dimensions, but there are many more classical and non-classical geometries besides. For another, more exotic example, one can consider the geometry of generic 2-plane distributions in 5 dimensions. Its flat model is ✕⇛, as a homogeneous space for the split real form of G_2 and the isomorphism ✕⇛ $\cong$ ✕—⇛, observed in Example (6.2.8) of this book, already has geometric consequences when fed through the parabolic machine. Although the details are fearsome, Nurowski [8] used this route to show that a generic 2-plane distribution in 5 dimensions gives rise to a conformal metric of signature $(3,2)$ on the same manifold.

More generally, phenomena on the flat model G/P are often reflected in the curved setting. For example, the flat connections on $\mathbb{E} = G \times_P E$ for G-modules $\mathbb{E}$ have curved versions known as *tractor connections*. These linear connections may be approached, and are in some sense equivalent to, the *Cartan connections* discussed in [2]. Here, the BGG machinery, and especially the incorporation of Lie algebra cohomology as in [3,4], provides the key, both to the construction of the Cartan connection, and to the theory of invariant differential operators [1].

Integral Geometry. Whilst parabolic differential geometry is a rather unexpected development in which some of the techniques in this book are useful only through serendipity, complex integral geometry was our main motivation. This subject has progressed substantially, as follows.

The classical Penrose transform concerns the analytic cohomology of the open orbit

$$\mathbf{PT}^+ \equiv \{L \subset \mathbb{C}^4 \mid \|\_\|^2|_L > 0\} \subset \mathbb{CP}_3$$

for the action of $\mathrm{SU}(2,2)$ preserving an Hermitian form $\|\_\|^2$ on $\mathbb{C}^4$ of signature $(2,2)$. Specifically, by means of the double fibration

$$\{L \subset P \subset \mathbb{C}^4\}$$

$$\eta \swarrow \qquad \searrow \tau$$

✕—●—● $= \mathbb{CP}_3 = \{L \subset \mathbb{C}^4\}$ $\qquad \{P \subset \mathbb{C}^4\} = \mathrm{Gr}_2(\mathbb{C}^4) = $ ●—✕—●

as discussed in §9.2 of this book, one may interpret $H^1(\mathbf{PT}^+, \mathcal{O}(k))$ as solutions of appropriate holomorphic differential equations on

$$\mathbf{M}^+ \equiv \{P \subset \mathbb{C}^4 \mid \|\_\|^2|_P \text{ is positive definite}\} \subset \mathrm{Gr}_2(\mathbb{C}^4).$$

More generally, one may consider any real form G_0 of a complex simple Lie group G and ask about its action on G/P for any parabolic subgroup $P \subset G$. If G_0 is compact, then such an action is transitive (and the analytic cohomology of G/P is the subject of the Bott–Borel–Weil Theorem). There are some rare cases when G_0 is non-compact but the action is transitive, e.g. $\mathrm{SL}(2, \mathbb{H})$ acting on $\mathbb{CP}_3$ (and the action of $\mathrm{SL}(2, \mathbb{H})$ on $\mathrm{Gr}_2(\mathbb{C}^4)$ is by conformal transformations on the 4-sphere as the unique minimal orbit (these observations forming the basis of 'Euclidean twistor theory')). Otherwise, there are always open orbits $\mathcal{D} \subsetneq G/P$—they are known as *flag domains* [7]. It is now known that flag domains enjoy the following features. Choose a maximal compact subgroup $K_0 \subset G_0$. It acts on $\mathcal{D}$ and there is a unique orbit $C_0 \subset \mathcal{D}$ that is a complex submanifold. For example, if $\mathrm{SU}(2,2) \subset \mathrm{SL}(4, \mathbb{C})$ is defined as preserving the Hermitian form

$$\|Z\|^2 = |Z_1|^2 + |Z_2|^2 - |Z_3|^2 - |Z_4|^2$$

then we can take

$$K_0 = \mathrm{S}(\mathrm{U}(2) \times \mathrm{U}(2)) = \left\{ \begin{pmatrix} * & * & 0 & 0 \\ * & * & 0 & 0 \\ 0 & 0 & * & * \\ 0 & 0 & * & * \end{pmatrix} \right\}$$

and $C_0 = \{[Z_1, Z_2, 0, 0]\} \subset \mathcal{D} = \mathbf{PT}^+$ is forced. In general we set

$$\mathcal{M}_\mathcal{D} \equiv \{gC_0 \mid g \in G \text{ and } gC_0 \subset \mathcal{D}\}_\circ,$$

the C_0-connected component of those G-translates of the base cycle C_0 that remain within $\mathcal{D}$. In our example, all translates of the projective line $C_0 \subset \mathbb{CP}_3$ make up the Grassmannian $\mathrm{Gr}_2(\mathbb{C}^4)$ and $\mathcal{M}_\mathcal{D} = \mathbf{M}^+$. In this way we are led to the double fibration as above. However, this case is deceptively misleading. Although, in general, the *cycle space* $\mathcal{M}_\mathcal{D}$ is naturally a Stein manifold, it is not usually itself a flag domain (nor even homogeneous under the action of G_0). Without going into detail, there is a trichotomy of flag domains [7, p. 57], namely

- the *Hermitian holomorphic* cases,
- the *Hermitian non-holomorphic* cases,
- the *non-Hermitian* cases,

typified by the following examples:

- $G_0 = \mathrm{SU}(2,2)$ acting on $\mathcal{D} = \mathbf{PT}^+ \subset \mathbb{CP}_3$,
- $G_0 = \mathrm{SU}(2,2)$ acting on $\mathcal{D} = \mathbf{M}^{+-} \subset \mathrm{Gr}_2(\mathbb{C}^4)$,
- $G_0 = \mathrm{SL}(4, \mathbb{R})$ acting on $\mathcal{D} = \mathbb{CP}_3 \setminus \mathbb{RP}_3 \subset \mathbb{CP}_3$,

where
$$\mathbf{M}^{+-} \equiv \{P \subset \mathbb{C}^4 \mid \|\_\|^2|_P \text{ is strictly indefinite}\}.$$

The cycle space of this particular flag domain may be identified as

$$\mathcal{M}_{\mathbf{M}^{+-}} = \mathbf{M}^+ \times \mathbf{M}^- \subset \mathrm{Gr}_2(\mathbb{C}^4) \times \mathrm{Gr}_2(\mathbb{C}^4)$$

but the associated double fibration is not part in this book. The cycles in $\mathbb{CP}_3 \setminus \mathbb{RP}_3$ are the non-singular quadrics with no real points. One can take as base cycle

$$C_0 = \{[Z_1, Z_2, Z_3, Z_4] \in \mathbb{CP}_3 \mid Z_1{}^2 + Z_2{}^2 + Z_3{}^2 + Z_4{}^2 = 0\}.$$

The cycle space is then realised as an open subset of $\mathrm{SL}(4, \mathbb{C})/\mathrm{SO}(4, \mathbb{C})$ and the associated double fibration is well outside this book. What is missing in both cases in a replacement for the efficiency of the BGG complex. More precisely, the 'Penrose transform in principle' still works as in Chapter 7 using the holomorphic relative de Rham complex along the fibres of η but the 'Penrose transform in practice,' as considered in Chapter 9, now has to do without the relative BGG complex.

  With hindsight, this book is about the double fibration transform for Hermitian holomorphic flag domains. Even the basic geometry of the other cases is much more difficult [7].

Finally, *real integral geometry* as typified by the classical Funk, Radon, and John transforms, bears a close resemblance to the complex case that is the subject of this book. Although a general theory is lacking, the complex methods of this book can be brought to bear to establish the kernel and range of certain real integral transforms.

  Here, suffice it to say that the classical John or 'X-ray' transform in three dimensions can be understood by means of the correspondence

$$\{(L \subset \mathbb{C}^4, P \subset \mathbb{R}^4) \mid \Re(L) \subseteq P\}$$

$$\eta \swarrow \qquad \qquad \searrow \tau$$

$$\mathbb{CP}_3 = \{L \subset \mathbb{C}^4\} \qquad \qquad \{P \subset \mathbb{R}^4\} = \mathrm{Gr}_2(\mathbb{R}^4)$$

(which is not a double fibration since η drops rank along $\mathbb{RP}_3 \subset \mathbb{CP}_3$). Details are in [6]—surely the beginnings of a more extensive theory.

Thanks. We are extremely grateful to Dover Publications for reprinting our book. Despite the original being typeset using LaTeX, it required some effort from Dover to enable our old files to compile. We have taken the opportunity to correct some mathematical misprints and we would also like to thank Susan Rattiner and Louise Jarvis for many corrections to our grammar. Finally, thanks for reading this new edition!

RoB and MikE, August 2016

Robert J. Baston
Previously a University Lecturer in Mathematics
at the Mathematical Institute and Wadham College, Oxford

Michael G. Eastwood
School of Mathematical Sciences
University of Adelaide

References for the Dover Preface

[1] R. J. Baston and M. G. Eastwood. Invariant operators. In: *Twistors in Mathematics and Physics*, Lond. Math. Soc. Lect. Notes **156**, Cambridge University Press 1990, pp. 129–163.

[2] A. Čap and J. Slovák. *Parabolic Geometries I: Background and General Theory.* Math. Surv. Monogr. **154**, Amer. Math. Soc. 2009.

[3] D. M. J. Calderbank and T. Diemer. Differential invariants and curved Bernstein–Gelfand–Gelfand sequences. *Jour. reine angew. Math.* **537** (2001), 67–103.

[4] A. Čap, J. Slovák, and V. Souček. Bernstein–Gelfand–Gelfand sequences. *Ann. Math.* **154** (2001), 97–113.

[5] M. G. Eastwood. A duality for homogeneous bundles on twistor space. *Jour. Lond. Math. Soc.* **31** (1985), 349–356.

[6] M. G. Eastwood. Complex methods in real integral geometry. *Suppl. Rendi. Circ. Mat. Palermo* **46** (1997), 55–71.

[7] G. Fels, A. Huckleberry, and J. A. Wolf. *Cycle Spaces of Flag Domains: a Complex Geometric Viewpoint.* Prog. Math. **245**, Birkhäuser 2006.

[8] P. Nurowski. Differential equations and conformal structures. *Jour. Geom. Phys.* **55** (2005), 19–49.

[9] Thales. The sphere is projectively flat. *Preprint*, Miletus, circa 600 BC.

PREFACE TO THE 1989 EDITION

One of the fundamental techniques in physics and mathematics is the discovery and exploitation of symmetry in a geometrical setting. The history of mathematics is filled with instances of its use, from Felix Klein's Erlangen programme, Sophus Lie's study of the invariants of differential equations and the work of Cartan and Weyl to the modern representation theory of reductive Lie groups in which the geometry of flag varieties plays a central rôle. In physics, there is the development of relativity and of modern quantum field theories in both of which the requirement is made that the equations of the theory exhibit natural invariances. Special relativity, for example, is the result of seeking a Poincaré invariant theory of mechanics; Einstein's motivation came from the Poincaré invariance of Maxwell's theory of electromagenetism.

Twistor theory, introduced by Roger Penrose in [116,118–120], is in part a geometrical study of *conformally* invariant physics in four spacetime dimensions. Maxwell's equations are conformally invariant as is a closely related family of equations, the *zero rest mass* field equations. This family describes massless fields of arbitrary spin; they include Weyl's equations for massless neutrinos and the equations of (anti-)self-dual linearized gravity. Initially, Penrose showed how solutions of these equations on spacetime could be expressed as contour integrals of free holomorphic functions over lines in three dimensional complex projective space; he later realized that the freedom allowed in the function for a fixed solution was exactly the freedom of a Čech representative of a sheaf cohomology class. The resulting isomorphism between a sheaf cohomology group on a region of projective space and solutions of a zero rest mass field equation on a region of spacetime has become known as the *Penrose transform*. Since his original work an entire industry has developed. Originally, the impetus came from mathematical physics and there is a wide literature from this point of view [90,120,121,126,128,155,157]. Additionally, much fruitful work has been done on the generalization of twistors and the Penrose transform to *curved* spaces [3,6,8,11,12,26,58,70,86,106,123,154].

The purpose of this monograph is to construct an analogue of the *Penrose transform* when the conformal group is replaced by any complex semisimple Lie group G and to study its relation to the representation theory of reductive Lie groups and algebras. Spacetime and projective space are replaced by suitable open subsets X and Z of compact complex homogeneous spaces or flag varieties for G; the transform calculates the cohomology of a homogeneous vector bundle on Z in terms of kernels and cokernels of invariant differential operators on sections of homogeneous vector bundles on X.

For the physicist, this means that we are able to give a systematic exploration of twistor theory in higher dimensions than four. There is much current physical interest in such generalizations since many mathematical models of our universe (e.g., Kaluza–Klein [148], supergravity [130], string theories [89], ...) view the usual four dimensional spacetime as a *dimensional reduction* of a rather larger manifold. Thus, we hope that the Penrose transform herein will find application in such settings. One such application has already been suggested by Witten [162]. We set up a simple algorithm for computing the transform so that our readers can explore the consequences of the theory for themselves. In doing this, we make use of two important results in representation theory, namely, Bernstein–Gelfand–Gelfand resolutions and the Bott–Borel–Weil theorem. We shall give proofs of these results which we hope physicists will find geometrically appealing.

For the mathematician, we wish to present the Penrose transform as of interest in its own right as a construction in representation theory. We shall see that it can be thought of as a geometric globalization of *Zuckerman's derived functor* construction. The transform can be used to construct unitary representations on cohomology groups (without the technical difficulties of L^2- cohomology) and to study homomorphisms between Verma modules (induced from a parabolic) especially those, called *non-standard*, associated to multiple subquotients of Verma modules.

Often, Z will be the complement of a *Schubert variety* and cohomology over Z will be isomorphic to cohomology supported on the variety (up to a finite dimensional error). Such relative cohomology occurs in the work of Beilinson–Bernstein as the cohomology of a $\mathcal{D}_\lambda$-module on the total flag variety G/B of G. Here, λ is a weight for the Lie algebra $\mathbf{g}$ of G. Via the Penrose transform, we can compute examples of such cohomology when λ is not dominant and see how the lack of dominance affects the structure of the resulting $\mathbf{g}$–modules. Non-standard homomorphisms of Verma modules seem to be easy to generate when we do this.

We have tried to write this book in a way which will be intelligible to both mathematicians and physicists. So we presuppose no acquaintance with twistor theory and a minimum of representation theory. This inevitably means that we shall have to spend time expounding some basic material in either camp and we apologize in advance for the irritation this may cause the experts. Since our aim is to use this material we have kept such expositions short—[94,127,128] will provide fuller explanations. We require a certain amount of mathematical sophistication ([25] contains most of the homological algebra we shall need, especially concerning spectral sequences) but we hope that in return the reader will be rewarded by the sight of abstract methods being used to compute concrete results. We are keen to give a practical "user's" guide to the transform and strongly encourage the reader to compute examples for each recipe encountered.

For completeness, we begin the book with an introductory chapter which sketches the development of the Penrose transform. In it we explain what Minkowski space is (for the non-physicist), show how solutions of the zero rest mass field equations can be given in terms of integral formulae, and interpret these geometrically. Along the way we note several topics (e.g., homogeneous spaces and vector bundles, cohomology, double fibrations) which we will take up in detail later. None of this chapter is needed in the remainder of the book so it may be omitted on a first reading. On the other hand, it could serve as a handy route map when the going gets tough.

Chapter 2 reviews the basic structure theory of complex semisimple Lie algebras and introduces the class of manifolds, namely complex homogeneous manifolds or flag varieties, on which the Penrose transform is defined. In it we introduce a useful notation, based on Dynkin diagrams, for representing parabolic subalgebras and the associated flag varieties, which we shall develop as the book advances until it (or, rather, the reader) is capable of computing the Penrose transform. In chapter 3 we review some representation theory, discuss homogeneous vector bundles, and adapt the diagram notation to these. Chapter 4 deals with the Weyl group of a semisimple Lie algebra and explicitly indicates its action on weights. This is in preparation for computing cohomology using the Bott–Borel–Weil theorem for which we give a simple proof (in the geometrical spirit of this book) in chapter 5. The theorem is the first main ingredient of the transform—we use it to compute direct images in the "push–down" stage. The reader who is already familiar with the theory of Lie algebras is encouraged to skim these chapters for notation.

Chapter 6 (which may be omitted on first reading) is by way of a digression; having developed the abstract theory of Lie algebras, Weyl groups, etc., we set ourselves an exercise on understanding a little more about the structure of flag varieties. We first see how they can be projectively embedded and given a cell decomposition. Then we compute their cohomology rings and, lastly, study their symplectic geometry. All this is well known, but it is derived easily at this point and is so beautiful that we could not resist its inclusion. We encourage the reader to skip this on a first reading and hurry along to the next chapter, which is where the book really begins.

Chapter 7 introduces the Penrose transform itself. We give it first in a very general setting of a complex or algebraic variety fibred over two varieties. The transform then relates cohomology on one of these to solutions of differential equations on the other. Later (in chapter 9) we specialize to double fibrations of the form

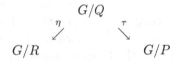

where G is a complex semisimple Lie group and P, Q and R are parabolic subgroups. We set $X \subset G/P$ to be open and, usually, Stein or affine, and $Z = \eta \circ \tau^{-1} X$. Then, if $\mathcal{E}$ is a homogeneous holomorphic vector bundle on Z, the Penrose transform computes $H^*(Z, \mathcal{E})$ in terms of invariant differential operators between sections of homogeneous sheaves on X.

The second main ingredient in computing the transform is a resolution, due to Bernstein–Gelfand–Gelfand, along the fibres of η. This is dealt with in chapter 8 and is based on the following simple idea. Suppose given on a manifold a family of vector fields X_i with the property that whilst they themselves may not span the tangent space at each point, the Lie algebra they generate does. Then a smooth function f is constant if and only if all $X_i f$ vanish. From this we can develop a differential resolution of the constants analogous to, but more efficient than, the de Rham resolution. Exactly this situation pertains on flag varieties and the construction yields the Bernstein–Gelfand–Gelfand resolutions.

All of this theory is brought together in chapter 9 where we are able to give many worked examples of the Penrose transform, from the standard four dimensional conformal Penrose twistor theory to higher dimensional twistor theories on conformal manifolds, on Grassmannians and for exceptional groups. Using the results of this chapter we show (chapter 10) how the Penrose transform can be used to construct realizations of unitary representations for real forms of G on the cohomology groups in the transform. We should stress again that this construction avoids L^2 cohomology and its technical difficulties. It is a cohomological translation of a physical inner product on solutions of invariant differential equations (specified by a Dirac operator, say).

Chapter 11 is devoted to understanding the various module structures on cohomology, comparing the Penrose transform with Zuckerman functors and with Beilinson–Bernstein theory. In it we study invariant differential operators between homogeneous sheaves and their relation to Verma modules and show how the Penrose transform, and therefore Zuckerman's functors, can be used to generate homomorphisms of Verma modules.

Chapter 12 reviews the constructions of the book and suggests further avenues for research.

It goes without saying that the inspiration for this work comes from Roger Penrose. We are grateful for the countless geometrical insights and other mathematics which he has provided. His influence and enthusiasm has been crucial and we hope that this comes through in what follows. The other major influence was John Rice who patiently and carefully explained a lot of representation theory to us and recognized the Bernstein–Gelfand–Gelfand resolution when it appeared in our work. This work would not exist without his help and encouragement. We would also like to thank Ed Dunne and Rod Gover for reading the manuscript, eliminating some (inevitably not all) of our errors, and making many useful comments.

Finally, we would like to thank our parents, for their encouragement.

CONTENTS

1

INTRODUCTION

In this first chapter we want to try to develop the Penrose transform from scratch. Penrose originally noticed that there are simple contour integral formulae[1] for solutions of a series of interesting equations from physics, namely the zero rest mass free field equations of mathematical physics. The Penrose transform is, in one sense, a machine for showing that all solutions can be obtained in this way. It is a lot more than that, for, as we shall see, it generates the equations as geometric invariants and says much about their symmetries.

Zero rest mass fields are of fundamental importance in physics. They describe electromagnetism, massless neutrinos, and linearized gravity. They all share the property that they are *conformally invariant*—that is they are determined by the conformal geometry of spacetime and so depend only on knowing how to measure relative lengths and angles, not on an overall length scale. Thus the equations which specify these fields are invariant under motions which preserve this conformal geometry and so the space of fields is invariant under the group of these motions. Our ultimate aim is to develop a Penrose transform when this group is any complex semisimple Lie group and to see what the representation theory of such groups implies about the Penrose transform.

We shall begin with a short study of Maxwell's equations for electromagnetism, which, in the absence of charges, amount to a pair of zero rest mass equations on Minkowski space. The idea is first to see where they are most naturally defined, and then to understand the contour integral formulae for them geometrically.

The rest of this book does not depend on an understanding of this chapter, which may therefore be omitted at a first reading. We hope, however,

---

[1]Similar to those of Bateman and Whitacker for the wave equation and Maxwell's equations [13].

that the reader will eventually take some time out to understand the origins
of the transform in mathematical physics.

Minkowski space

Maxwell's equations for a free electromagnetic field (i.e., one in the absence
of charges) can be succinctly written down using differential forms as
follows:

$$dF = 0 \quad \text{and} \quad d{*}F = 0.$$

Here F is a two-form on $\mathbf{R}^4$ which represents the electric and magnetic
fields, d is exterior differentiation, and $*$ is the Hodge star operation with
respect to a (flat) Lorentz metric g on $\mathbf{R}^4$. With indices, they may be
written as

$$\nabla_{[a}F_{bc]} = 0 \quad \text{and} \quad \nabla_{[a}{*}F_{bc]} = 0,$$

where $[\dots]$ indicates that the enclosed indices are to be skewed over. Here
$*F_{ab} = \frac{1}{2}\epsilon_{ab}{}^{cd}F_{cd}$ where ϵ_{abcd} is the volume form with respect to g and we
are using the Einstein summation convention (as in [127], for example).
It is easy to check from these formulae that $*$ acting on two–forms is independent of the scale of the metric g—any metric $\kappa^2 g$ yields the same $*$
on two-forms. Of course, d is invariant under any diffeomorphism of $\mathbf{R}^4$,
and it follows that Maxwell's equations are invariant under the *conformal*
motions of $\mathbf{R}^4$, i.e., diffeomorphisms of $\mathbf{R}^4$ which preserve g up to scale.
These include the Poincaré motions which are those globally defined con-
formal motions which preserve the scale of g. Now on two forms, $*^2 = -1$
and so we may write

$$F = \phi + \tilde{\phi}$$

where ϕ and $\tilde{\phi}$ are in the $\pm i$ eigenspaces of $*$—in particular, they are necessarily complex two-forms (and conjugate). In terms of these, Maxwell's
equations become

$$d\phi = 0 \quad \text{and} \quad d\tilde{\phi} = 0.$$

There are three observations to make here. First, if the metric g is
taken to be Euclidean, then $*^2 = 1$ and there is no need to introduce complex two-forms to obtain the decomposition; on the other hand, we may
simply choose to allow F to take complex values and replace $\mathbf{R}^4$ by $\mathbf{C}^4$.
This turns out to be a very convenient thing to do and even the most natural thing to do when we study contour integral formulae, in a moment.
It may seem rather strange physically, but even then it is a wise move,
especially if we have quantum mechanics in mind—there we are actually
interested in the analytic continuation of fields from $\mathbf{R}^4$ to tube domains

in $\mathbf{C}^4$. There is also the bonus of not having to distinguish between different signatures for g. So, from now on, we work over $\mathbf{C}$ and refer to four dimensional complex Euclidean space (with its flat holomorphic metric) as *affine Minkowski space*.

The second observation is that not all conformal diffeomorphisms of affine Minkowksi space are well defined everywhere. For example, if $\mathbf{x}, \mathbf{b} \in \mathbf{C}^4$ then the mapping

$$\mathbf{x} \longmapsto \mathbf{y} = \frac{\mathbf{x} + \mathbf{b}\|\mathbf{x}\|^2}{1 + 2\mathbf{b} \cdot \mathbf{x} + \|\mathbf{x}\|^2 \|\mathbf{b}\|^2}$$

is conformal but not defined on the light cone of the point $-\mathbf{b}/\|\mathbf{b}\|^2$. To rectify this we need to compactify affine Minkowski space. This is very similar to forming the Riemann sphere from $\mathbf{C}$ so that fractional linear (Möbius) transforms are globally defined. We will do this in detail in a later chapter; it turns out that the right choice is the Grassmannian $\mathbf{Gr}_2(\mathbf{C}^4)$ of all two dimensional subspaces of $\mathbf{C}^4$. The embedding is given by sending the point $\mathbf{x} = (x^0, x^1, x^2, x^3)$ of affine Minkowski space to the subspace spanned by the vectors

$$\begin{pmatrix} 1 \\ 0 \\ x^0 + x^1 \\ x^2 - ix^3 \end{pmatrix} \quad \text{and} \quad \begin{pmatrix} 0 \\ 1 \\ x^2 + ix^3 \\ x^0 - x^1 \end{pmatrix}. \tag{1}$$

Denote the image of $\mathbf{x}$ in $\mathbf{Gr}_2(\mathbf{C}^4)$ by x. We shall refer to the resulting conformal compactification as *Minkowski space*. Then the global conformal motions of Minkowski space can be realized as the group $\mathrm{SL}(4,\mathbf{C})$ (modulo its centre) with its natural action. Indeed, consider the matrix

$$\begin{pmatrix} 1 & 0 & b^0 - b^1 & -b^2 - ib^3 \\ 0 & 1 & -b^2 + ib^3 & b^0 + b^1 \\ 0 & 0 & 1 & 0 \\ 0 & 0 & 0 & 1 \end{pmatrix}.$$

Applying this to the two vectors representing $\mathbf{x}$ yields the image of $\mathbf{y}$ in $\mathbf{Gr}_2(\mathbf{C}^4)$.

Homogeneous bundles on Minkowski space

This brings us to our third observation—the two-form F is a section of a *homogeneous vector bundle* on Minkowski space and the decomposition given above is its reduction into its irreducible components. Let us briefly recall the natural bundles on $\mathbf{Gr}_2(\mathbf{C}^4)$. The simplest is the so–called *tautological* bundle, S', whose fibre S'_x at x is the two dimensional subspace $x \subset \mathbf{C}^4$ itself. Similarly, there is the *quotient* bundle S whose fibre S_x at x is $\mathbf{C}^4/x$. S', S are the *spinor* bundles on Minkowski space. In Penrose's abstract index notation [44,127] they are denoted

$$S' = \mathcal{O}_{A'} \quad \text{and} \quad S = \mathcal{O}^A.$$

Their duals are denoted

$$S'^* = \mathcal{O}^{A'} \quad \text{and} \quad S^* = \mathcal{O}_A.$$

(Thus the natural pairing between dual bundles is achieved by contraction between an upper and lower index.) Both S and S' are *homogeneous* bundles—this means that the action of $\mathrm{SL}(4,\mathbf{C})$ on Minkowski space lifts to an action on sections of these bundles. This is easy to see, since any element of $\mathrm{SL}(4,\mathbf{C})$ mapping x to y in Minkowski space is by definition a map sending S'_x to S'_y. Furthermore, both bundles are *irreducible* in the sense that the isotropy group of x acts irreducibly on S_x, S'_x. Put another way, neither contains a proper homogeneous subbundle. Bundles formed from S, S' by taking tensor products, direct sums, etc., are also homogeneous, in the obvious way.

Now it is a standard fact that on any Grassmannian the tangent bundle is the tensor product of the quotient bundle and the dual of the tautological bundle; so the tangent bundle of Minkowski space is

$$\Theta = S \otimes S'^* = \mathcal{O}^{AA'}$$

and the cotangent bundle, or bundle of one–forms, is

$$\Omega^1 = S^* \otimes S' = \mathcal{O}_{AA'}.$$

From this it is easy to compute that two-forms are sections of

$$\Omega^2 = [\odot^2 S^* \otimes \wedge^2 S'] \oplus [\wedge^2 S^* \otimes \odot^2 S'],$$

where $\odot^k S'$ indicates the kth symmetric power of S'. The bundle $L = \wedge^2 S$ is called the *determinant* line bundle on Minkowski space. It is convenient to fix an element of $\wedge^4 \mathbf{C}^4$ so that we can identify $L = \wedge^2 S'^*$. In the notation of [44] $L = \mathcal{O}[1]$ and $L^* = \wedge^2 S^* = \mathcal{O}[-1]$. Then

$$\Omega^2 = \mathcal{O}_{(AB)}[-1] \oplus \mathcal{O}_{(A'B')}[-1]$$

(where $(A'B')$ indicates that the enclosed indices are to be symmetrized). This gives the decomposition $F = \phi + \tilde{\phi}$. Maxwell's equations become

$$\nabla^A_{A'}\phi_{AB} = 0 \quad \text{and} \quad \nabla^{A'}_A \tilde{\phi}_{A'B'} = 0.$$

To write these equations we have had to choose a metric locally on Minkowski space and form the Levi–Civita connection $\nabla_{AA'}$ on spinors. We must choose a metric in the *conformal class* of metrics. To see what this means, notice that a metric must be a section of

$$\odot^2 \Omega^1 = [\odot^2 S^* \otimes \odot^2 S'] \oplus [\wedge^2 S^* \otimes \wedge^2 S'] \tag{2}$$

and is in the conformal class if its projection onto the first factor is zero. Such a metric must have the form

$$g_{ab} = \epsilon_{AB}\tilde{\epsilon}_{A'B'}$$

where ϵ_{AB} and $\tilde{\epsilon}_{A'B'}$ are antisymmetric and each is a square root of g_{ab} (noting $\wedge^2 S^* \cong \wedge^2 S'$). Let ϵ^{AB} and $\tilde{\epsilon}^{A'B'}$ be their inverses. $\nabla_{AA'}$ is defined by the requirement that it be torsion free and preserve both ϵ's; define $\nabla^A_{A'} = \epsilon^{AB}\nabla_{BA'}$ and $\nabla^{A'}_A$ similarly. The fact that Maxwell's equations are conformally invariant corresponds to the fact that the operators

$$\phi_{AB} \to \nabla^A_{A'}\phi_{AB} \quad \text{and} \quad \tilde{\phi}_{A'B'} \to \nabla^{A'}_A \tilde{\phi}_{A'B'}$$

do *not* depend on the metric g, but only on the decomposition (2).

Penrose's contour integrals

Consider the second of these equations. Penrose has given a contour integral formula for its solutions [119,128]:

$$\tilde{\phi}_{A'B'}(x) = \oint \pi_{A'}\pi_{B'}f(\eta(\pi,x))\pi^{D'}d\pi_{D'}.$$

To interpret this formula, let x be fixed and let $\pi_{A'} \in S'_x$ determine a vector $z = \eta(\pi,x) \in \mathbf{C}^4$. f is a holomorphic function on an appropriate region of $\mathbf{C}^4$ which is homogeneous of degree -4 so that $f(\lambda z) = \lambda^{-4}f(z)$ for $\lambda \in \mathbf{C}$. Let $\pi^{D'} = \epsilon^{D'E'}\pi_{E'}$. It is easy to see that the integrand is independent of the scale of π (because of the homogeneity of f.) It is therefore well defined on a domain contained in the projective line $\mathbf{P}S'_x$ of one dimensional subspaces of x. By requiring f to have appropriately situated singularities, we may suppose that this domain is not simply connected and choose to evaluate the integral over a non-trivial contour. We may also suppose that this prescription may be carried out smoothly as we vary x.

To check that $\tilde{\phi}$ is a solution we confine ourselves to *affine* Minkowksi space X. Then $\nabla_{AA'} = \partial/\partial x^{AA'}$, where $x^{AA'}$ is the matrix

$$
\begin{pmatrix}
x^0 + x^1 & x^2 + ix^3 \\
x^2 - ix^3 & x^0 - x^1
\end{pmatrix}
$$

and the index A labels rows whilst A' labels columns. Following (1), x is the column span of

$$
\begin{pmatrix}
I^{C'}_{A'} \\
X^{AC'}
\end{pmatrix}
$$

and

$$
z = \eta(\pi, x) = \begin{pmatrix} I^{C'}_{A'} \\ x^{AC'} \end{pmatrix} \pi_{C'} = \begin{pmatrix} \pi_{A'} \\ x^{AC'} \pi_{C'} \end{pmatrix}.
$$

It follows that we may write

$$
\frac{\partial}{\partial x^{CC'}} f(z) = \pi_{C'} f_C(z),
$$

and so

$$
\nabla^{A'}_C \tilde{\phi}_{A'B'}(x) = \epsilon^{A'C'} \nabla_{C'C} \tilde{\phi}_{A'B'}
$$
$$
= \oint \epsilon^{A'C'} \pi_{A'} \pi_{B'} \pi_{C'} f_C(z) \pi^{D'} \, d\pi_{D'}
$$
$$
= 0
$$

(by the antisymmetry of $\epsilon^{A'C'}$).

We required that f be homogeneous of degree -4; it is clear from this calculation that if f is homogeneous of degree $-n - 2$ and if

$$
\psi_{A'B'...C'}(x) = \oint \underbrace{\pi_{A'} \pi_{B'} \ldots \pi_{C'}}_{n \text{ terms}} f(\eta(\pi, x)) \pi^{D'} \, d\pi_{D'},
$$

then

$$
\nabla^{A'}_A \psi_{A'B'...C'} = 0,
$$

where $\psi \in \odot^n S' \otimes L^* = \mathcal{O}_{(A'B'...C')}[-1]$. Solutions of these equations are called *free zero rest mass* fields of helicity $n/2$. The operator

$$
\psi_{A'B'...C'} \longmapsto \nabla^{A'}_A \psi_{A'B'...C'}
$$

is conformally invariant and is nothing more than the *Dirac–Weyl operator* of helicity $n/2$.

Let us try to give a geometrical interpretation of these integrals. Notice that f should be thought of as a section of a homogeneous line bundle over $\mathbf{CP}^3$. The natural homogeneous line bundle on $\mathbf{CP}^3$ is again a *tautological* bundle, H, whose fibre at a point is the one dimensional subspace of $\mathbf{C}^4$ specified by that point. We claim that if f is homogeneous of degree -1 on $\mathbf{C}^4$ then f determines a section of H. This is easy to see: for $\mathbf{Z} \in \mathbf{C}^4$ consider $f(\mathbf{Z})\mathbf{Z} \in \mathbf{C}^4$. This is constant along any one dimensional subspace of $\mathbf{C}^4$ and defines an element of that subspace. Moreover, if f is homogeneous of degree $-k < 0$ then f defines a section of $H^{\otimes k}$. (This bundle is usually denoted by $\mathcal{O}(-k)$.)

Notice also that f may not be defined over all of $\mathbf{CP}^3$, for otherwise, by Cauchy's theorem, ψ would be zero. To avoid this, f should have singularities arranged in such a way that $f(\eta(\pi, x))$ is non-singular over, say, an annulus in $\mathbf{P}S'_x$ for each $x \in X$. There is also a natural freedom to change f without affecting ψ; by Cauchy's theorem, again, if we add to f a function $\hat{f}$ with $\hat{f}(\eta(\pi, x))$ non-singular over a *contractible* region of $\mathbf{P}S'_x$ for each $x \in X$, ψ is unchanged.

Penrose recognized that this freedom is exactly the freedom of a Čech representative of a *first cohomology class* with values in $\mathcal{O}(-n-2)$ defined over the region Z in $\mathbf{CP}^3$ swept out by all the lines $\mathbf{P}S'_x$. So contour integration gives a map

$$\mathsf{P} : H^1(Z, \mathcal{O}(-n-2)) \to \text{zero rest mass fields } \psi_{A'B'\ldots C'} \text{ on } X.$$

(It has to be checked that the domain of P is all of the cohomology group.)

We shall see that P is an *isomorphism*. The general machine which has been developed to prove this is the *Penrose transform*. We can construct this machine from the contour integral formula by trying to give it a geometrical interpretation.

The Penrose transform

The first thing to note is that the contour integral is really defined on the projective bundle $\mathbf{P}S'$, for that is where its integrand lives. The second is that it is well defined if we replace $f(\eta(\pi, x))$ by any function f' of x and π which is homogeneous of degree $-n-2$ in π (although, of course, the resulting field ψ may not then satisfy the zero rest mass equations). What does such a function represent? The reader will be unsurprised to learn that it is a section of a homogeneous line bundle. To see this, observe that

$$\mathbf{P}S' = \{F_1 \subset F_2 \subset \mathbf{C}^4 \mid \dim F_i = i\} = \mathbf{F}_{12}$$

is a partial flag variety with an evident transitive action of $\mathrm{SL}(4,\mathbf{C})$; if H' is the tautological line bundle assigning to $([\pi], x)$ the span of π, then f' gives a section of $H'^{\otimes n+2}$. Adapting the comments of the previous paragraph we see that, provided the singularities of f' are appropriately situated, f' represents a first cohomology class with values in H'. If $\tau : \mathbf{P}S' \to \mathbf{Gr}_2(\mathbf{C}^4)$ sends $([\pi], x)$ to x and if $Y = \tau^{-1}X$, then

$$f' \in H^1(Y, H'^{\otimes n+2}).$$

The point is that contour integration identifies

$$H^1(Y, H'^{\otimes n+2}) \cong \Gamma(X, \mathcal{O}_{A'B'\ldots C'}[-1]).$$

Put another way, cohomology on Y can be computed by first computing cohomology along the fibres of $Y \to X$ (using a contour integral) and then by taking sections over X.

Now let us understand where the zero rest mass equations arise. If η denotes the map $\mathbf{P}S' \to \mathbf{CP}^3$, then it is evident that $H' = \eta^* H$ and that any section f of H defines a section $\eta^* f$ of H' by the formula

$$\eta^* f(\pi, x) = f(\eta(\pi, x)).$$

This means there is a map

$$\eta^* \; : \; H^1(Z, \mathcal{O}(-k)) \to H^1(Y, H'^{\otimes k}). \tag{3}$$

Furthermore, any f' is of the form $\eta^* f$ only if it is constant along the fibres of η; we can express this as a differential equation:

$$f'_A = \pi^{A'} \nabla_{AA'} f'(\pi, x) = 0.$$

f'_A is also a section of a homogeneous vector bundle on $\mathbf{P}S'$; we shall simply call this $\Omega_\eta(H')$ to indicate that it arises when we differentiate sections of H' along the fibres of η. (The justification for the notation will come later.) Contour integration (for example) shows that

$$H^1(Y, \Omega_\eta(H'^{\otimes n+2})) \cong \Gamma(X, \mathcal{O}_{A\underbrace{(B'\ldots C')}_{n-1 \text{ terms}}}[-2])$$

and that $\pi^{A'} \nabla_{AA'}$ induces the operator

$$H^1(Y, H'^{\otimes n+2}) \;\to\; H^1(Y, \Omega_\eta(H'^{\otimes n+2}))$$
$$\psi_{A'B'\ldots C'} \;\longmapsto\; \nabla^{A'}_A \psi_{A'B'\ldots C'}.$$

The facts that $\eta^* H^1(Z, \mathcal{O}(-k-2))$ is the kernel of this operator and that η^* is an isomorphism onto this kernel now seems intuitively reasonable (although there is a substantial amount of technical detail to be checked here).

We may summarise this discussion as follows. We have been considering a *double fibration* or *correspondence*

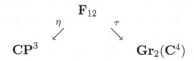

of generalized flag varieties. Given an open (affine) $X \subset \mathbf{Gr}_2(\mathbf{C}^4)$, such as affine Minkowski space, we induce a sub-double fibration

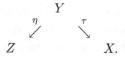

Using η^*, we identify cohomology on Z with coefficients in a homogeneous sheaf $\mathcal{F}$ with cohomology on Y of the sheaf $\eta^{-1}\mathcal{F}$ (an isomorphism requiring a topological restriction on X, in general—affinity is sufficient). This is a submodule of the cohomology of $\eta^*\mathcal{F}$ on Y, distinguished by means of certain invariant differential equations along the fibre of η. Finally, we interpret the cohomology of $\eta^*\mathcal{F}$ and these differential equations in terms of sections of homogeneous bundles on X. Modulo technical questions, this should express $H^*(Z, \mathcal{F})$ in terms of invariant differential equations on X.

This, in a nutshell, is the Penrose transform. We have glossed over several technical matters and swept a large amount of homological algebra (or *abstract nonsense* as some call it) into a future chapter where it properly belongs. But at last we are in a position to explain to the reader what this book is really about.

The Penrose transform on flag varieties

Our aim is to provide the reader with a streamlined and finely tuned version of this machine in the context of arbitrary complex semisimple Lie groups. To construct it we shall replace $\mathbf{Gr}_2(\mathbf{C}^4)$ and $\mathbf{CP}^3$ by two compact complex homogeneous manifolds G/P and G/R for an arbitrary complex semisimple Lie group G. P, Q are *parabolic* subgroups of G and we shall require that $Q = P \cap R$ is parabolic also. The basic geometric setting is the double fibration

$$G/Q$$
$$\eta \swarrow \quad \searrow \tau$$
$$G/R \qquad\qquad G/P.$$

Let X be an open submanifold of G/P; X may, for example, be a maximal affine subvariety (an affine "big cell") or a union of open orbits of a real form of G. Set $Y = \tau^{-1}X$ and $Z = \eta X$. Let $\mathcal{F}$ be the sheaf of germs of holomorphic or regular sections of a homogeneous vector bundle on G/R. By following the ideas sketched above, we shall interpret $H^i(Z, \mathcal{F})$ in terms of kernels and cokernels of invariant differential operators on sections of homogeneous vector bundles on X. The invariant differential equations on Y will be induced by *Bernstein–Gelfand–Gelfand resolutions* and the cohomology of the homogeneous sheaves on Y will be computed as sections of homogeneous sheaves on X using the *Bott–Borel–Weil* theorem. The whole construction will be manifestly invariant under the action of the Lie algebra $\mathbf{g}$ of G and under any subgroup of G which preserves X.

So now it is time to begin our journey. We do so with a review of the theory of Lie algebras and their representations.

2

LIE ALGEBRAS AND FLAG MANIFOLDS

This and the next few chapters are concerned with developing the mathematical background we shall need for the Penrose transform. They consist of (brief) reviews of the representation theory of complex semisimple Lie algebras and the geometry of *flag manifolds*.

In this chapter, we review the structure and geometry of *generalized flag manifolds*; these are the varieties of parabolic subalgebras of complex semisimple Lie algebras and so are homogeneous spaces G/P where G is a complex semisimple Lie group and P is a parabolic subgroup. As complex manifolds, they are compact, Kähler (indeed, projective), and simply connected; they exhaust all such manifolds which also admit a holomorphic homogeneous action of a complex Lie group.

There is no loss of generality in assuming that this Lie group G is simply connected, in which case its properties are determined by the Lie algebra **g** associated to G. Accordingly, we review the structure theory of semisimple complex Lie algebras and their parabolic subalgebras, introducing a particularly useful notation, using Dynkin diagrams, to distinguish parabolic subalgebras and flag manifolds. In fact, we shall usually assume that our Lie algebra **g** is simple, the semisimple case being a fairly trivial extension of the simple theory.

Several examples are constructed: in particular, we give a brief exposition of *fibrations* of generalized flag manifolds.

The reader who is unfamiliar with the details of the structure theory of Lie algebras may wish to consult [94] and more detail on generalized flag manifolds may be found in [18,85,156].

2.1 Some structure theory

Let **g** be a complex semisimple Lie algebra and G the associated simply connected Lie group. A maximally Abelian self-normalizing subalgebra **h** is called a *Cartan subalgebra*, and any two of these are conjugate under the adjoint action of G. Fix such a subalgebra, once and for all, and consider its adjoint action on $\mathbf{g}$:

$$\text{for } h \in \mathbf{h}, \; v \in \mathbf{g}, \; (\operatorname{ad} h)v = [h, v].$$

Then $\mathbf{g}$ decomposes as a direct sum of joint eigenspaces under this action. The zero eigenspace is $\mathbf{h}$ itself. All others (called root spaces) are one-dimensional. We set, for $\alpha \in \mathbf{h}^* \equiv \operatorname{Hom}_{\mathbf{C}}(\mathbf{h}, \mathbf{C})$,

$$\mathbf{g}_\alpha = \{v \in \mathbf{g} \text{ s.t. } [h, v] = \alpha(h)v\}.$$

The collection

$$\Delta(\mathbf{g}, \mathbf{h}) = \{\alpha \in \mathbf{h}^* \text{ s.t. } \mathbf{g}_\alpha \neq \{0\}, \alpha \neq 0\}$$

encodes much of the structure of $\mathbf{g}$ and is called the set of *roots* of $\mathbf{g}$ relative to $\mathbf{h}$. (Having fixed $\mathbf{h}$, we shall often simply write Δ for $\Delta(\mathbf{g}, \mathbf{h})$.) Thus,

$$\mathbf{g} = \mathbf{h} \oplus \bigoplus_{\alpha \in \Delta} \mathbf{g}_\alpha.$$

Since, by the Jacobi identity on a Lie algebra, $[\mathbf{g}_\alpha, \mathbf{g}_\beta] \subseteq \mathbf{g}_{\alpha+\beta}$ (which may be zero), $\Delta(\mathbf{g}, \mathbf{h})$ spans an integral lattice in $\mathbf{h}^*$ called the *root lattice*: let $\mathbf{h}_{\mathbf{R}}^*$ be the real span of that lattice. Then $\mathbf{h}_{\mathbf{R}}^*$ is totally real and has real dimension the same as the complex dimension of $\mathbf{h}^*$. In other words, $\mathbf{h}^*$ is the complexification of $\mathbf{h}_{\mathbf{R}}^*$.

A subset $\mathcal{S} \subset \Delta$ with the property that every $\alpha \in \Delta$ may be expressed as a linear combination of elements of $\mathcal{S}$ with all non-negative *or* all non-positive coefficients is called a *system of simple roots* of $\mathbf{g}$. Such $\mathcal{S}$ exist and then $\mathcal{S}$ is a basis for $\mathbf{h}^*$ and Δ is contained in the integral lattice generated by $\mathcal{S}$. Any two such $\mathcal{S}$ are conjugate. We shall fix a choice of $\mathcal{S}$ once and for all. There is then an induced partial ordering on $\mathbf{h}^*$; if $\lambda, \mu \in \mathbf{h}^*$, write

$$\lambda \succeq \mu \iff \lambda - \mu = \sum_i a_i \alpha_i \text{ with } \alpha_i \in \mathcal{S} \text{ and } a_i \geq 0.$$

The subset

$$\Delta^+(\mathbf{g}, \mathbf{h}, \mathcal{S}) \equiv \{\alpha \in \Delta(\mathbf{g}, \mathbf{h}) \text{ s.t. } \alpha \succ 0 \text{ w.r.t. } \mathcal{S}\}$$

is called the set of *positive roots* (with respect to $\mathcal{S}$).

Now $\mathbf{g}$ admits a bilinear form, the *Killing form*, as follows: if $u, v \in \mathbf{g}$, let $\operatorname{ad} u$ and $\operatorname{ad} v$ be the corresponding endomorphisms of $\mathbf{g}$ given by

$$(\operatorname{ad} u)x = [u, x], \text{ etc.}$$

Then set

$$\langle u, v \rangle = \operatorname{tr}(\operatorname{ad} u)(\operatorname{ad} v).$$

This bilinear form is non-degenerate on $\mathbf{h}$ and, hence on $\mathbf{h}^*$, which is thus identified with $\mathbf{h}$ by means of $\langle\ ,\ \rangle$. It restricts to a positive definite quadratic form on $\mathbf{h}_{\mathbf{R}}^*$.

It is the central theorem of the structure theory of complex semisimple Lie algebras that a knowledge of the conformal structure of this Euclidean space together with the basis $\mathcal{S}$ determines and is determined by $\mathbf{g}$. Specifically, if $\alpha \in \Delta(\mathbf{g}, \mathbf{h})$ and

$$\alpha^{\vee} = 2\alpha/\langle \alpha, \alpha \rangle$$

is its *co-root*, then for $\mathcal{S} = \{\alpha_i\}$, the *Cartan integers*

$$c_{ij} = \langle \alpha_i, \alpha_j^{\vee} \rangle,$$

called the *Cartan matrix*, uniquely specify $\mathbf{g}$. This matrix is severely restricted. A useful shorthand to describe it is a so-called *Dynkin diagram* (as in table 2.1). This is a graph (with some directed edges) whose nodes correspond to the simple roots α_i and whose edges determine c_{ij}. Thus:

1. $\langle \alpha_i, \alpha_i^{\vee} \rangle = 2$

2. $\alpha_i \neq \alpha_j$ are connected *iff* $\langle \alpha_i, \alpha_j^{\vee} \rangle \neq 0$

3. $\overset{\alpha\ \ \ \beta}{\bullet\!\!-\!\!\bullet} \iff \langle \alpha, \beta^{\vee} \rangle = -1$

   $\overset{\alpha\ \ \ \beta}{\bullet\!\!\Rightarrow\!\!\bullet} \iff \langle \alpha, \beta^{\vee} \rangle = -2 \quad \langle \beta, \alpha^{\vee} \rangle = -1$

   $\overset{\alpha\ \ \ \beta}{\bullet\!\!\Rrightarrow\!\!\bullet} \iff \langle \alpha, \beta^{\vee} \rangle = -3 \quad \langle \beta, \alpha^{\vee} \rangle = -1.$

2.2 Borel and parabolic subalgebras

We shall be concerned with a certain class of subalgebras of $\mathbf{g}$. The simplest of these are the Borel subalgebras defined to be the maximal solvable subalgebras of $\mathbf{g}$. The prototype or *standard* Borel subalgebra is

$$\mathbf{b} = \mathbf{h} \oplus \mathbf{n}$$

Table 2.1. Dynkin diagrams of simple Lie algebras

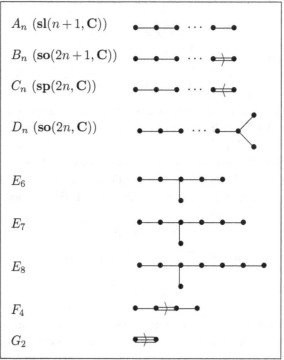

where

$$\mathbf{n} = \bigoplus_{\alpha \in \Delta^+(\mathbf{g},\mathbf{h})} \mathbf{g}_\alpha.$$

Example (2.2.1).   If $\mathbf{g} = \mathbf{sl}(n, \mathbf{C})$, then $\mathbf{h}$ consists of the diagonal matrices, $\mathbf{n}$ the strictly upper triangular matrices, and $\mathbf{b}$ the upper triangular matrices.

Every Borel subalgebra is G-conjugate to the standard prototype $\mathbf{b}$. A parabolic subalgebra of $\mathbf{g}$ is one which contains a Borel subalgebra. Again, there exists a standard form for these. Given $\mathbf{g}$, $\mathbf{h}$, and $\mathcal{S}$ as usual, let $\mathcal{S}_\mathbf{p}$ be a subset of $\mathcal{S}$ and set

$$\Delta(\mathbf{l}, \mathbf{h}) = \operatorname{span} \mathcal{S}_\mathbf{p} \cap \Delta(\mathbf{g}, \mathbf{h})$$

so that

$$\mathbf{l} = \mathbf{h} \oplus \bigoplus_{\alpha \in \Delta(\mathbf{l},\mathbf{h})} \mathbf{g}_\alpha$$

is a subalgebra of $\mathbf{g}$. $\mathbf{l}$ is reductive: $\mathbf{l} = [\mathbf{l}, \mathbf{l}] + \mathbf{l}_Z$ where $\mathbf{l}_Z$ is the centre of $\mathbf{l}$ and $\mathbf{l}_S = [\mathbf{l}, \mathbf{l}]$ is semisimple of rank $= |\mathcal{S}_{\mathbf{p}}|$. Let $\Delta(\mathbf{u}, \mathbf{h}) = \Delta^+(\mathbf{g}, \mathbf{h}) \setminus \Delta(\mathbf{l}, \mathbf{h})$ and take

$$\mathbf{u} = \bigoplus_{\alpha \in \Delta(\mathbf{u}, \mathbf{h})} \mathbf{g}_\alpha.$$

Then $\mathbf{u}$ is nilpotent, and the subalgebra

$$\mathbf{p} = \mathbf{l} \oplus \mathbf{u} \tag{4}$$

contains $\mathbf{b}$ and is hence parabolic. Up to conjugation, all parabolics arise in this way. The decomposition (4) of $\mathbf{p}$ is called a *Levi* decomposition. It will also be useful to set

$$\mathbf{u}_- = \bigoplus_{\alpha \in \Delta(\mathbf{u}, \mathbf{h})} \mathbf{g}_{-\alpha}.$$

This gives a complement to $\mathbf{p}$ in $\mathbf{g}$, that is a subalgebra $\mathbf{u}_-$ with

$$\mathbf{g} = \mathbf{u}_- \oplus \mathbf{p}.$$

A parabolic is called *maximal* if it is a maximal proper subalgebra of $\mathbf{g}$. Then $\mathcal{S}_{\mathbf{p}}$ is just one element short of $\mathcal{S}$.

A useful notation for a standard parabolic $\mathbf{p} \subseteq \mathbf{g}$ is to cross through all nodes in the Dynkin diagram for $\mathbf{g}$ which correspond to the simple roots of $\mathbf{g}$ in $\mathcal{S} \setminus \mathcal{S}_{\mathbf{p}}$. For example, the standard Borel subalgebra has crosses through *every* node whilst a maximal parabolic has only *one* crossed node. The Levi factor $\mathbf{l}$ of $\mathbf{p}$ then comprises a semisimple part $\mathbf{l}_S$, whose Dynkin diagram is obtained from that of $\mathbf{p}$ by deleting crossed nodes and incident edges, and a centre $\mathbf{l}_Z \subseteq \mathbf{h}$ of dimension $|\mathcal{S} \setminus \mathcal{S}_{\mathbf{p}}| =$ number of crossed nodes. We may choose a Cartan subalgebra $\mathbf{h}_{\mathbf{l}}$ of $\mathbf{l}_S$ so that $\mathbf{h}_{\mathbf{l}} \oplus \mathbf{l}_Z = \mathbf{h}$. Specifically, for our standard parabolic, if we use the Killing form to identify $\mathbf{h}$ with $\mathbf{h}^*$, then we should take

$$\mathbf{h}_{\mathbf{l}} = \operatorname{span} \mathcal{S}_{\mathbf{p}} = \mathbf{h} \cap \mathbf{l}_S.$$

Example (2.2.2).   Let $\mathbf{g} = \mathrm{sl}(4, \mathbf{C})$ with $\mathbf{h}$ and $\mathcal{S}$ as above. It can readily be checked that, for example,

$$\bullet\!\!-\!\!\times\!\!-\!\!\bullet = \left\{ \begin{pmatrix} * & * & * & * \\ * & * & * & * \\ \hline 0 & 0 & * & * \\ 0 & 0 & * & * \end{pmatrix} \in \mathrm{sl}(4, \mathbf{C}) \right\}$$

$$\times\!\!-\!\!\bullet\!\!-\!\!\bullet = \left\{ \begin{pmatrix} * & * & * & * \\ 0 & * & * & * \\ 0 & * & * & * \\ 0 & * & * & * \end{pmatrix} \in \mathbf{sl}(4,\mathbf{C}) \right\}.$$

The subalgebras $\mathbf{l}$, $\mathbf{u}$ consist, respectively, of the "diagonal blocks" and the "top right hand block."

2.3   Generalized flag varieties

Again, let $\mathbf{g}$ be a complex semisimple Lie algebra and $\mathbf{p} \subseteq \mathbf{g}$ a fixed parabolic subalgebra. Let G be the associated simply connected Lie group with $\mathbf{g}$ as Lie algebra. Then $\mathbf{p}$ gives rise to a subgroup of G which will be denoted by P. The subgroups which arise in this way are called the *parabolic* subgroups of G. All are conjugate to a *standard* parabolic subgroup, i.e., one obtained from a standard parabolic subalgebra. Let X be the set of all parabolics in G which are conjugate to P. The stabilizer of the conjugation action of G on X is exactly P whence X may be regarded as a homogeneous space:

$$X = G/P$$

as in, for example, [95,139]. As pointed out above, X is simply connected and compact. In fact, P is parabolic if and only if G/P is compact [95]. As a complex manifold it is Kähler; indeed we will show how to construct a projective embedding in section 6.1. Conversely, Wang [152] has shown that all compact simply connected homogeneous Kähler manifolds are of this form.

These spaces will be called *generalized flag varieties*, in honour of the case $\mathbf{g} = \mathbf{sl}(n,\mathbf{C})$. To designate a particular generalized flag manifold we shall abuse notation and denote X by the same Dynkin diagram as $\mathbf{p}$.

Example (2.3.1).   The basic examples of flag manifolds are the varieties of complete flags in a vector space V; a complete flag in V is a nested sequence of subspaces

$$V_1 \subset V_2 \subset \ldots \subset V_m = V,$$

where $\dim V_i = i$. If $\{e_i\}$ is a basis of V, then a standard flag is given by $V_i = \mathrm{span}\{e_1, e_2, \ldots, e_i\}$. $SL(V)$ clearly acts on this variety, with stabilizer the subgroup B of upper triangular matrices, at the standard flag. Thus,

$$\mathbf{F}_c(V) = \text{Variety of complete flags in } V = SL(V)/B.$$

In the notation introduced above, $\mathbf{F}_c = \times\!\!-\!\!\times\!\!-\!\!\times \cdots \times\!\!-\!\!\times$ ($m-1$ nodes). The simplest case is $\times = \mathbf{CP}_1$.

More generally, we may consider the varieties of partial flags in V. Let $\mathbf{n} = (a, b, \ldots, d)$ be a strictly increasing sequence of natural numbers with $d < m = \dim V$; then a partial flag of type $\mathbf{n}$ in V is a sequence of nested subspaces

$$V_a \subset V_b \subset \ldots \subset V_d \subset V,$$

where $\dim V_e = e$. A standard flag is easily constructed from a basis $\{e_i\}$ and, relative to this basis, the stabilizer of the standard flag is the parabolic

$$\mathbf{p} = \underset{}{\bullet}\!\!-\!\!\underset{}{\bullet} \ \cdots \ \underset{}{\bullet}\overset{\alpha_a}{\underset{\times}{}}\underset{}{\bullet} \ \cdots \ \underset{}{\bullet}\overset{\alpha_b}{\underset{\times}{}}\underset{}{\bullet} \ \cdots, \quad \text{etc.,}$$

realizing the variety $\dot{\mathbf{F}}_{\mathbf{n}}(V)$ of partial flags of type $\mathbf{n}$ in V as a generalized flag variety. Projective spaces and Grassmannians are particular examples of such varieties.

Example (2.3.2). In Penrose's *twistor theory* [90,126,128,155], we should take $G = \mathrm{SL}(4, \mathbf{C})$. To see why this is so, begin with complex (affine) Minkowski space. It is just $\mathbf{C}^4$ with a complex metric, which may be written in *null coordinates* as

$$\|\mathbf{y}\|^2 = 2(y_1 y_4 - y_2 y_3),$$

where $\mathbf{y} \in \mathbf{C}^4$. Similarly, if $\mathbf{x} \in \mathbf{C}^6$ write

$$\|\mathbf{x}\|^2 = 2(x_1 x_6 - x_2 x_5 + x_3 x_4).$$

Then define a mapping $\mathbf{C}^4 \to \mathbf{C}^6$ by

$$\phi : \mathbf{y} \to \left(\frac{1}{2}, y_1, \ldots, y_4, \|\mathbf{y}\|^2\right)$$

so that $\|\phi(\mathbf{y})\| = 0$, i.e., $\phi(\mathbf{y})$ is *null*.

Twistor theory, however, concerns itself primarily with the *conformal* geometry of Minkowski space; so we should allow for the replacement of the metric on $\mathbf{C}^4$ by a conformally rescaled metric

$$\|\mathbf{y}\|_\kappa^2 = 2\kappa^2(y_1 y_4 - y_2 y_3)$$

with $\kappa \in \mathbf{C}$ and adjust ϕ to

$$\phi_\kappa : \mathbf{y} \to \left(\frac{\kappa}{2}, \kappa y_1, \ldots, \kappa y_4, \kappa^{-1}\|\mathbf{y}\|_\kappa^2\right)$$

so that again $\|\phi_\kappa(\mathbf{y})\| = 0$. Of course, $\phi_\kappa = \kappa\phi$ so that the conformal geometry of $\mathbf{C}^4$ defines an embedding of affine Minkowski space into $\mathbf{CP}_5$ whose image is evidently an open dense subvariety of the quadric corresponding to the null cone of the origin in $\mathbf{C}^6$. In twistor theory, it is customary to call this quadric (complexified, compactified) *Minkowski space*; it is a conformal manifold, for locally we may use ϕ_κ to define a complex metric up to scale. Of course, this quadric is also the complexification of the four sphere (since signature is irrelevant over $\mathbf{C}$.) Denote it by

$$\mathbf{M} = \mathbf{CS}^4.$$

Spin$(6, \mathbf{C})$ acts transitively on $\mathbf{M}$ and a moment's thought shows that it covers the group of conformal motions of M which is just the image of Spin$(6, \mathbf{C})$ in PSL$(6, \mathbf{C})$ so that the covering is $4 - 1$. Now recall that Spin$(6, \mathbf{C})$ is just SL$(4, \mathbf{C})$, for we may identify $\mathbf{C}^6$ with $\wedge^2 \mathbf{C}^4$ and SL$(4, \mathbf{C})$ is simply connected. The metric on $\mathbf{C}^6$ is defined by the following: let $\eta \in \wedge^4 \mathbf{C}^4$ be a volume form preserved by SL$(4, \mathbf{C})$ and, if $\mathbf{x} \in \wedge^2 \mathbf{C}^4$, then

$$x \wedge x = \|\mathbf{x}\|^2 \eta.$$

So a vector is null if and only if it is represented by a *simple* two-form; but simple two forms defined up to scale are in one-to-one correspondence with two dimensional subspaces of $\mathbf{C}^4$ so that

$$\mathbf{M} = \mathrm{Gr}_2(\mathbf{C}^4) = \bullet\!\!-\!\!\times\!\!-\!\!\bullet\,.$$

As an even dimensional quadric, $\mathbf{M}$ has two distinct families of (totally null) two-planes usually called *α-planes* and *β-planes*. These families constitute generalized flag varieties also: they are projective spaces

$$\mathbf{PT} = \times\!\!-\!\!\bullet\!\!-\!\!\bullet \quad \text{and} \quad \mathbf{PT}^* = \bullet\!\!-\!\!\bullet\!\!-\!\!\times.$$

Each point in $\mathbf{M}$ corresponds to a complex line in $\mathbf{PT}$ or $\mathbf{PT}^*$—a fact classically known as the *Klein correspondence*. Any two α-planes or β-planes in $\mathbf{M}$ intersect in a single point. However, an α-plane and a β-plane generically miss each other. If they happen to intersect, then they intersect in a line which is null in the conformal structure on $\mathbf{M}$. All null lines of $\mathbf{M}$ arise in this way and so the space of such lines is the partial flag variety

$$\mathbf{N} = \times\!\!-\!\!\bullet\!\!-\!\!\times.$$

Remark (2.3.3).   In standard twistor theory, $\mathbf{N}$ is called *ambitwistor space* $\mathbf{A}$, since a point of $\mathbf{A}$ is a pair of points $([Z^\alpha], [W_\alpha])$ (in homogeneous coordinates) with $[Z^\alpha] \in \mathbf{PT}$, $[W_\alpha] \in \mathbf{PT}^*$, and $Z^\alpha W_\alpha = 0$. This extremely useful situation has no analogue for spaces of null lines in dimensions other than four (see example 2.3.5).

The remaining partial flag varieties of $\mathbf{T}$ play the rôle of correspondence spaces in the Penrose transform.

Remark (2.3.4).   The observations of example 2.3.2 led one of us [50,51] to generalize the well-understood Penrose transform between these spaces [44] to a transform between all partial flag varieties; this book may be regarded as a continuation of that programme to generalized flag varieties.

Example (2.3.5).   It is easy to generalize example 2.3.2 to higher dimensions; the analogue of Minkowski space will be the complexified sphere or complex quadric $\mathbf{CS^m}$. Simply replace $\mathbf{C}^4$ and $\mathbf{C}^6$ by $\mathbf{C}^m$ and $\mathbf{C}^{m+2}$ and adapt the formulae for the metrics and the embeddings ϕ_k in the obvious way.

In even dimensions $m = 2n$, we find

$$\mathbf{CS^{2n}} = \text{×———•———•} \cdots \text{•——} \Big\langle \quad (n+1 \text{ nodes}).$$

$\mathbf{CS^{2n}}$ again contains two distinct families of n-planes, namely

$$\mathbf{Z^{2n}} = \text{•———•———•} \cdots \text{•——} \Big\langle^{×}$$

and

$$\mathbf{Z^{*2n}} = \text{•———•———•} \cdots \text{•——} \Big\langle_{×}$$

each being $n(n+1)/2$-dimensional. They generalize the (projective) twistor spaces for $\mathbf{M}$ and so we shall call them *higher dimensional twistor spaces*. In example 6.1.3 we shall identify them as the projective spaces of reduced pure spinors for $SO(2n,\mathbf{C})$.

In odd dimensions $m = 2n+1$,

$$\mathbf{CS^{2n+1}} = \text{×———•———•} \cdots \text{•}\!\Rightarrow\!\!\text{•} \quad (n+1 \text{ nodes}).$$

Odd-dimensional quadrics have only *one* family of (totally null) n-planes. This is the generalized flag variety

$$\mathbf{Z^{2n+1}} = \text{•———•———•} \cdots \text{•}\!\Rightarrow\!\!\text{×}$$

of dimension $(n+1)(n+2)/2$. It is again the natural generalization of the twistor space for $\mathbf{M}$ and the projective space of pure $SO(2n+1,\mathbf{C})$ spinors.

In even dimensions, there is a natural analogue of the ambitwistor space in four dimensions, namely

$$\mathbf{A}^{2n} = \quad \bullet\!-\!\!-\!\bullet\!-\!\!-\!\bullet \ \cdots \ \bullet\!-\!\!-\!\!<\begin{smallmatrix}\times\\ \\ \times\end{smallmatrix} \quad .$$

However, for $n \neq 2$, $\mathbf{A}^{2n}$ is not the space of null lines in $\mathbf{CS}^{2n}$. This latter space $\mathbf{N}^{2n}$ may be identified as follows: LeBrun observed [105] that the space of null geodesics in any conformal manifold admits a contact structure. The homogeneous contact manifolds were identified by Boothby [21], there being exactly one for each complex simple Lie algebra. So

$$\mathbf{N}^{2n} = \quad \bullet\!-\!\!\times\!\!-\!\bullet \ \cdots \ \bullet\!-\!\!-\!\!<\begin{smallmatrix}\bullet\\ \\ \bullet\end{smallmatrix}$$

$$\mathbf{N}^{2n+1} = \quad \bullet\!-\!\!\times\!\!-\!\bullet \ \cdots \ \bullet\!\!\Rightarrow\!\!\bullet \quad .$$

Of course, these identifications may also be made by pure thought.

2.4   Fibrations of generalized flag varieties

If $\mathbf{q} \subset \mathbf{p}$ are parabolic subalgebras of $\mathbf{g}$, then we have a natural fibration

$$G/Q \xrightarrow{\tau} G/P.$$

The fibre is easily identified: let $\mathbf{p} = \mathbf{l} \oplus \mathbf{u}$ be a Levi decomposition with L_S the semisimple simply connected Lie group corresponding to the semisimple part $\mathbf{l}_S$ of the reductive factor $\mathbf{l}$. Then $\mathbf{q} \cap \mathbf{l}_S$ is parabolic in $\mathbf{l}_S$ and the fibre of τ is isomorphic to $P/Q \cong L/(L \cap Q) \cong L_S/(L_S \cap Q)$ and, consequently, a generalized flag manifold. The Dynkin diagram for the fibre is evidently obtained from the following

---

Recipe (2.4.1).   Delete from the Dynkin diagram for $\mathbf{q}$ all crossed nodes (and incident edges) *shared* with $\mathbf{p}$ and then delete all connected components with no crossed through nodes.

Thus, for example, the fibration

$$\mathbf{F}_{123} = \ \times\!-\!\!\times\!-\!\!\times \ \rightarrow \ \bullet\!-\!\!\times\!-\!\bullet \ = \mathbf{M}$$

has fibres isomorphic to

$$\times \ \ \times \ = \mathbf{CP}^1 \times \mathbf{CP}^1.$$

Similarly, the fibration

$$\mathbf{G^{2n}} = \ \ \longrightarrow \ \ = \mathbf{CS^{2n}}$$

has fibre

$$\text{(one node fewer)} = \mathbf{CS^{2n-2}}$$

Taking $\mathbf{p}$ and $\mathbf{r}$ to be standard parabolics, now otherwise unrestricted, $\mathbf{p} \cap \mathbf{r} = \mathbf{q}$ is a standard parabolic (with Dynkin diagram obtained by crossing through all nodes which are crossed in either of $\mathbf{p}$ or $\mathbf{r}$). Thus, to any pair of generalized flag varieties, G/P and G/R, there is associated a double fibration:

$$
\begin{array}{ccc}
 & G/Q & \\
\eta \swarrow & & \searrow \tau \\
G/R & & G/P.
\end{array}
$$

Notice that τ and η provide an embedding

$$G/Q \hookrightarrow G/P \times G/R$$

so that the fibres of τ are embedded into G/R by means of η and vice versa. This gives a *correspondence* (see page 73) between points of G/P and certain submanifolds of G/R and vice versa. This is exactly the situation obtaining between Minkowski space and twistor space in Penrose's original construction.

Example (2.4.2). The double fibration

$$\mathbf{G^{2n}} = $$

$$
\begin{array}{ccc}
\eta \swarrow & & \searrow \tau
\end{array}
$$

$$\mathbf{N^{2n}} = \qquad\qquad\qquad\qquad\qquad\qquad = \mathbf{CS^{2n}}$$

has $\times$ for the fibre of η; thus, points of $\mathbf{N}^{2n}$ correspond to lines in $\mathbf{CS}^{2n}$. On the other hand, as we have seen, the fibre of τ is $\mathbf{CS}^{2n-2}$, so all the lines in $\mathbf{CS}^{2n}$ (arising from points in $\mathbf{N}^{2n}$) through a particular point in $\mathbf{CS}^{2n}$ generate the null cone through that point, as expected.

So now we have the basic geometric structure required in Penrose's original transform. We next need a class of bundles on these spaces, whose cohomology is to be the subject of the transform.

HOMOGENEOUS VECTOR BUNDLES ON G/P

In this chapter we recall the construction of homogeneous vector bundles on homogeneous spaces G/P from representations of P. We shall use as our building blocks those which arise from irreducible representations of P and, occasionally, the restriction to P of an irreducible representation of G. Thus, we must briefly review the representation theory of semisimple Lie algebras; in particular, we introduce a useful notation for weights and representations with which we shall be able to compute later. We also briefly discuss the question of taking pull-backs of homogeneous bundles under homogeneous fibrations, since this forms the start of the Penrose transform.

3.1 A brief review of representation theory

In section 2.1 we fixed, once and for all, a Cartan subalgebra $\mathbf{h}$ and a set of simple roots S of $\mathbf{g}$. Recall that a *weight* for $\mathbf{g}$ is an element of $\mathbf{h}^*$ and that, if $v \in V$ with V a representation space of $\mathbf{g}$, then v is a weight vector for $\mathbf{h}$, of weight $\lambda \in \mathbf{h}^*$, if and only if

$$\forall h \in \mathbf{h}, \ hv = \lambda(h)v.$$

The set of all vectors in V of weight λ is called the λ-*weight space* of V, its dimension being the *multiplicity* of λ. Denote by $\Delta(V)$ the set of all weights whose weight spaces in V are non-trivial. If $\lambda \in \Delta(V)$, then λ is said to be a weight of V. Let

$$\mathbf{n} = \bigoplus_{\alpha \in \Delta^+(\mathbf{g},\mathbf{h})} \mathbf{g}_\alpha ; \quad \mathbf{n}_- = \bigoplus_{\alpha \in \Delta^+(\mathbf{g},\mathbf{h})} \mathbf{g}_{-\alpha}$$

be the *raising* and *lowering* subalgebras of $\mathbf{g}$. A vector $v \in V$ is called *maximal* (resp. *minimal*) if and only if it is annihilated by $\mathbf{n}$ (resp. $\mathbf{n}_-$)

under the action of $\mathbf{g}$. Recall that $\mathbf{h}^*$ has a partial ordering as in section 2.1; then a maximal (resp. minimal) vector of weight λ is *highest* in V (resp. *lowest* in V) if $\lambda \succeq \lambda'$ (resp. $\lambda \preceq \lambda'$) $\forall \lambda' \in \Delta(V)$.

Next, consider the weights in $\mathbf{h}_\mathbf{R}^*$. The form $\langle\ ,\ \rangle$ on $\mathbf{h}^*$ restricts to a positive definite form on $\mathbf{h}_\mathbf{R}^*$. For each $\alpha \in \Delta(\mathbf{g}, \mathbf{h})$, W_α is the *wall* or hyperplane in $\mathbf{h}_\mathbf{R}^*$ perpendicular to α. These walls partition $\mathbf{h}_\mathbf{R}^*$ into distinct open regions called *Weyl chambers*. A unique chamber, the *fundamental chamber*, is distinguished by requiring its elements λ to satisfy

$$\langle \lambda, \alpha \rangle > 0 \quad \forall \alpha \in \Delta^+(\mathbf{g}, \mathbf{h}, \mathcal{S})$$

(equivalently, $\forall \alpha \in \mathcal{S}$). A weight in the closure of the fundamental chamber is called *dominant* for $\mathbf{g}$. A weight not on any wall is called *regular*.

Given $\mathcal{S} = \{\alpha_j\}$, define a dual set of weights $\{\lambda_i\}$ by requiring

$$\langle \lambda_i, \alpha_j^\vee \rangle = \delta_{ij}.$$

Then $\{\lambda_i\}$ is a basis for $\mathbf{h}^*$ and any $\lambda \in \mathbf{h}^*$ is expressed as

$$\lambda = \sum_i \langle \lambda, \alpha_i^\vee \rangle \lambda_i.$$

λ is said to be *integral* if and only if the coefficients in this sum are integers; clearly, λ is dominant if and only if they are non-negative. We need a

---

Notation for weights:
Represent a weight λ for $\mathbf{g}$ by inscribing the coefficient $\langle \lambda, \alpha_j^\vee \rangle$ over the jth node of the Dynkin diagram for $\mathbf{g}$.

Example (3.1.1).

The classification theorem of irreducible finite dimensional representations of semisimple Lie algebras, $\mathbf{g}$, is:

Theorem (3.1.2).   *A finite dimensional irreducible representation of $\mathbf{g}$ has a unique highest weight vector, of weight λ, which is dominant integral for $\mathbf{g}$, and this induces a one-to-one correspondence between finite dimensional irreducible $\mathbf{g}$-modules and dominant integral weights.*

If V is such a representation space, then V^*, its dual, has a lowest weight vector of weight $-\lambda$. So the theorem may equivalently be given in terms of *lowest weights*. We shall adopt this seemingly perverse point of view because the Bott–Borel–Weil theorem (which we shall use extensively below) is most easily expressed in this form. Furthermore, we shall extend our abuse of notation and allow the Dynkin diagram for λ to denote also the representation with lowest weight $-\lambda$. (This is consistent with the notation we shall shortly establish for vector bundles. View a representation of $\mathbf{g}$ as a vector bundle over $G/G = \{\mathrm{pt}\}$.)

If $\mathbf{p}$ is a parabolic subalgebra of $\mathbf{g}$, then, since $\mathbf{h} \subset \mathbf{p}$, weights have a meaning also for $\mathbf{p}$. Notice that if $\mathbf{p} = \mathbf{l} \oplus \mathbf{u}$ is a Levi decomposition of $\mathbf{p}$, then $\mathbf{u}$ acts trivially on any irreducible representation of $\mathbf{p}$ (since, by Engel's theorem, $\mathbf{u}$ acts by nilpotent endomorphisms). Thus, an irreducible representation of $\mathbf{p}$ corresponds to an irreducible representation of $\mathbf{l}$. Let $\tau : \mathbf{h}^* \to (\mathbf{h} \cap \mathbf{l}_S)^*$ be the natural projection. Then a weight λ is *integral* or *dominant* for $\mathbf{p}$ according as $\tau(\lambda)$ is integral or dominant for $\mathbf{l}_S$. These are the obvious conditions on the nodal coefficients $\langle \lambda, \alpha_i^\vee \rangle$ for $\alpha_i \in \mathcal{S}_{\mathbf{p}}$. Namely, coefficients over the uncrossed nodes are constrained to be integral or non-negative, respectively. Then we obtain

Theorem (3.1.3).   *The finite dimensional irreducible representations of $\mathbf{p}$ are in one-to-one correspondence with $\lambda \in \mathbf{h}^*$ which are dominant and integral for $\mathbf{p}$.*

Remark (3.1.4).   A similar comment concerning lowest weights is applicable here and the Dynkin diagram for λ (with choice of $\mathbf{p}$ recorded by means of crosses) will be used to denote the representation of $\mathbf{p}$ whose lowest weight is $-\lambda$.

Remark (3.1.5).   Since the same weights could be used to determine representations for different parabolics $\mathbf{p}$ and $\mathbf{p}'$, this distinction must be incorporated into the notation for the corresponding representation. This is automatic in the Dynkin diagram notation. A useful alternative is to write $F_{\mathbf{p}}(\mu)$ for the irreducible representation of $\mathbf{p}$ with μ as an extremal weight. Thus, $F_{\mathbf{p}}(-\lambda)$ is the required representation of $\mathbf{p}$. In case $\mathbf{p} = \mathbf{g}$, we shall simply write $F(-\lambda)$. These notations agree with [151] but we shall further set

$$E_{\mathbf{p}}(\lambda) \equiv F_{\mathbf{p}}(-\lambda) \quad \text{and} \quad E(\lambda) \equiv F(-\lambda).$$

Again, our reasons for these slightly odd conventions stem from the Bott–Borel–Weil theorem of chapter 5 and also our use of the Bernstein–Gelfand–Gelfand resolution of chapter 8 in a form dual to the usual.

Remark (3.1.6).   In fact, we are really concerned with Lie groups and their representations rather than just Lie algebras. For semisimple $\mathbf{g}$, the classification coincides with that for the associated simply connected Lie group G. However, if the parabolic $\mathbf{p} \subset \mathbf{g}$ corresponds to $P \subset G$, then in order for an irreducible representation of $\mathbf{p}$ to "exponentiate" to one for P, it is necessary and sufficient that the dominant weight λ in theorem 3.1.3 above be integral for $\mathbf{g}$ and not just for $\mathbf{p}$. We shall assume this to be the case unless otherwise indicated.

Example (3.1.7).   Recall that any irreducible finite dimensional representation of $\mathbf{sl}(n+1, \mathbf{C})$ is obtained as an irreducible submodule of the tensor algebra on the self-representation of $\mathbf{sl}(n+1, \mathbf{C})$. Indeed, a similar statement is true for $\mathrm{GL}(n+1, \mathbf{C})$ and the representations of $\mathbf{sl}(n+1, \mathbf{C})$ are obtained by restricting to $\mathrm{SL}(n+1, \mathbf{C})$ and differentiating. The irreducible finite dimensional representations of $\mathrm{GL}(n+1, \mathbf{C})$ may be specified by means of a *Young diagram* as in table 3.1 [50,112,127,159]. If the abstract index approach is used, as in [127], then $\mathbf{C}^{n+1} = V^{\alpha}$ and the Young diagrams with m boxes specify symmetries on the m-fold tensor power of V^{α}:

$$\bigotimes^{m}\mathbf{C} = V^{\overbrace{\alpha\beta\cdots\delta}^{m}}.$$

Strictly speaking a Young diagram does rather more than just this. The imposition of symmetries may be regarded as projecting from $V^{\alpha\beta\cdots\delta}$ onto a copy of the required representation by a $\mathrm{GL}(n+1, \mathbf{C})$-equivariant homomorphism. Since such representations are apt to occur inside $V^{\alpha\beta\cdots\delta}$ with multiplicity greater than one there is a good deal of choice for such a

Table 3.1.   A Young diagram

projection. Indeed, the symmetric group $\mathcal{S}_m$ acts transitively on the space of such projections via its action on $V^{\alpha\beta\cdots\delta}$. This action provides an irreducible representation of $\mathcal{S}_m$ and all such representations of $\mathcal{S}_m$ arise in this way. In other words, Young diagrams also parametrize the irreducible representations of $\mathcal{S}_m$ and are best thought of as describing the splitting of $V^{\alpha\beta\cdots\delta}$ under the action of $GL(n+1, \mathbf{C}) \times \mathcal{S}_m$.

For any Young diagram thought of as a representation of $\mathcal{S}_m$ a basis for the representation space is provided by filling in the m boxes with the numbers $1, 2, \ldots, m$ such that each row and each column is in increasing order, forming a *Young tableaux*. Thus, $(1, 1)$ is a two dimensional representation of $\mathcal{S}_3$ with basis

1	2
3	

1	3
2	

These two elements describe, in the notation of [127], the particular projections

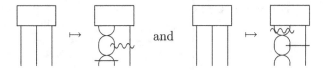

Restricting to $SL(n+1, \mathbf{C})$ eliminates the determinant representation. In other words, under $SL(n+1, \mathbf{C})$ we have

$$(a, b, \ldots, d) \cong (a+n, b+n, \ldots, d+n)$$

for any n. This redundancy shows up in our notation for the finite dimensional irreducible representations of $\mathbf{sl}(n+1, \mathbf{C})$:

$$(a, b, c, d, \ldots, e, f, g) = \overset{b\text{-}a}{\bullet}\!\!-\!\!\overset{c\text{-}b}{\bullet}\!\!-\!\!\overset{d\text{-}c}{\bullet} \quad \cdots \quad \overset{f\text{-}e}{\bullet}\!\!-\!\!\overset{g\text{-}f}{\bullet}$$

so that, in particular, the self representation is

$$\overset{0}{\bullet}\!\!-\!\!\overset{0}{\bullet}\!\!-\!\!\overset{0}{\bullet} \quad \cdots \quad \overset{0}{\bullet}\!\!-\!\!\overset{1}{\bullet}.$$

(The reader will easily check that this is consistent with the "dual" approach that we are taking when specifying representations using Dynkin diagrams.)

Thus, for the mathematical physicist, (non-projective) twistor spaces are

$$\mathbf{T}^{\alpha} = \overset{0}{\bullet}\!\!-\!\!\overset{0}{\bullet}\!\!-\!\!\overset{1}{\bullet} \qquad\qquad \mathbf{T}_{\alpha} = \overset{1}{\bullet}\!\!-\!\!\overset{0}{\bullet}\!\!-\!\!\overset{0}{\bullet}$$

$$\mathbf{T}_{(\alpha\beta)} = \overset{2}{\bullet}\!\!-\!\!\overset{0}{\bullet}\!\!-\!\!\overset{0}{\bullet} \qquad\qquad \mathbf{T}_{[\alpha\beta]} = \overset{0}{\bullet}\!\!-\!\!\overset{1}{\bullet}\!\!-\!\!\overset{0}{\bullet} = \mathbf{T}^{[\alpha\beta]}.$$

Note that $\overset{0}{\bullet}\!\!-\!\!\overset{1}{\bullet}\!\!-\!\!\overset{0}{\bullet}$ may also be regarded as the self-representation of $\mathbf{so}(6,\mathbf{C})$ and the last of these demonstrates the isomorphism of this self-representation with its dual, afforded by the metric preserved by $\mathbf{so}(6,\mathbf{C})$.

Example (3.1.8). Spinors are of great importance in mathematical physics; these correspond to the following representations (cf., e.g., [81,149]):

For SO$(2n, \mathbf{C})$:
$$\begin{cases} \overset{0}{\bullet}\!\!-\!\!\overset{0}{\bullet}\!\!-\!\!\overset{0}{\bullet} \cdots \overset{0}{\bullet}\!\!-\!\!\overset{0}{\bullet}\!\!\!\overset{\displaystyle 1}{\diagup}_{\displaystyle 0} \\[2em] \overset{0}{\bullet}\!\!-\!\!\overset{0}{\bullet}\!\!-\!\!\overset{0}{\bullet} \cdots \overset{0}{\bullet}\!\!-\!\!\overset{0}{\bullet}\!\!\!\overset{\displaystyle 0}{\diagup}_{\displaystyle 1} \end{cases}$$

For SO$(2n+1, \mathbf{C})$: $\overset{0}{\bullet}\!\!-\!\!\overset{0}{\bullet}\!\!-\!\!\overset{0}{\bullet} \cdots \overset{0}{\bullet}\!\!\Rightarrow\!\!\overset{1}{\bullet}$.

Example (3.1.9). The adjoint representation of SL$(3,\mathbf{C})$ is $\overset{1}{\bullet}\!\!-\!\!\overset{1}{\bullet}$. The "baryon octet," consisting of spin-$\frac{1}{2}$ particles, the proton, the neutron, and their siblings, is regarded as providing a basis for this representation (of the corresponding real form SU(3)), as described, for example, in [111]. This symmetry group is known as a *gauge group*, the classification scheme being known as the "eight-fold way" of Gell-Mann.

Example (3.1.10). An important representation of a parabolic $\mathbf{p} \subset \mathbf{g}$ is the adjoint action of $\mathbf{p}$ on $\mathbf{g}/\mathbf{p}$. Unless $\mathbf{p}$ is a maximal parabolic, it is evident that the representation is reducible (we see from the root space decomposition of $\mathbf{g}$ that $\mathbf{g}/\mathbf{p}$ has no highest weight vector). Maximality is a necessary but not sufficient condition for irreducibility. To investigate further (following Kobayashi and Nagano [100]) let

$$\mathbf{p} = \mathbf{l} \oplus \mathbf{u} \qquad\qquad \mathbf{g} = \mathbf{u}_- \oplus \mathbf{p}$$

be fixed with $\mathcal{S}_{\mathbf{p}} = \mathcal{S} \setminus \{\alpha\}$. Fix $e \in \mathbf{l}_{\mathbb{Z}}$ by $\alpha(e) = 1$ and observe that the eigenvalues of $\operatorname{ad} e$ are integral, negative on $\mathbf{u}_-$, zero on $\mathbf{l}$, and positive on $\mathbf{u}$. If $\mathbf{p}$ acts irreducibly on $\mathbf{g}/\mathbf{p}$, then the eigenvalues of $\operatorname{ad} e$ are -1, 0, or 1; conversely, if this is so, then, by a simple argument involving *roots strings* [94], $\mathbf{p}$ acts irreducibly on $\mathbf{g}/\mathbf{p}$. Equivalently, the coefficient of α in an expansion of the highest root θ of $\mathbf{g}$ (in terms of $\mathcal{S}$) is 1. When $\mathbf{g}/\mathbf{p}$ is irreducible, the lowest weight of $\mathbf{g}$, namely $-\theta$, is the lowest weight of this representation; by consulting a table of highest root vectors for simple $\mathbf{g}$ (e.g., [94]), we can list the irreducible $\mathbf{g}/\mathbf{p}$ as in table 3.2. The general structure of the representation is explained in section 9.9.

Remark (3.1.11).   For $\mathbf{p} \subset \mathbf{g}$ as above, the decomposition $\mathbf{g} = \mathbf{u}_- \oplus \mathbf{l} \oplus \mathbf{u}$ is a $|1|$-*grading of* $\mathbf{g}$ (cf. [115]); this is significant for the

Table 3.2.   Irreducible $\mathbf{g}/\mathbf{p}$

extension of the Penrose transform (and its various invariant differential operators) to *curved spaces* [8,11].

Remark (3.1.12).   Note that $\mathbf{g}/\mathbf{p}$ is irreducible if and only if $\mathbf{u}_-$ is Abelian (this latter condition being equivalent to ad e having eigenvalues in $\{-1, 0, 1\}$).

3.2   Homogeneous bundles on G/P

We can now construct the bundles which will be the subject of the Penrose transform in this book. Let $\mu : P \to \mathrm{End}(V)$ be a representation of P on a finite dimensional $\mathbf{C}$-vector space V and induce the *homogeneous vector bundle*

$$\mathcal{V} = G \times_P V$$
$$\downarrow$$
$$G/P.$$

A section of $\mathcal{V}$ may be thought of as a V-valued function on G satisfying the relation

$$f(gp) = \mu(p^{-1})f(g)$$

(for $g \in G$ and $p \in P$). We shall usually take f to be holomorphic and so an equivalent differential condition is (for $x \in \mathbf{p}$, thought of as a *left* invariant vector field on G):

$$x[f] = -d\mu(x)f.$$

The sheaf of germs of holomorphic sections of $\mathcal{V}$ is denoted by $\mathcal{O}(V)$; if V is the trivial representation on $\mathbf{C}$, then $\mathcal{O}(V) = \mathcal{O}$, the holomorphic structure sheaf of G/P. Thus, for each integral weight for $\mathbf{g}$ which is dominant for $\mathbf{p}$ one obtains an irreducible representation of P (as in remark 3.1.6) and hence an irreducible homogeneous bundle on G/P. In this case we shall write $\mathcal{O}_{\mathbf{p}}(\lambda)$ for $\mathcal{O}(E_{\mathbf{p}}(\lambda))$. With the Dynkin diagram notation, we can omit the reference to $\mathbf{p}$. Thus,

$$\mathcal{O}(\overset{p}{\underset{\bullet}{}} \overset{q}{\underset{\times}{\rule{1cm}{0.4pt}}} \overset{r}{\underset{\bullet}{}}) \quad (p, r \geq 0)$$

is a bundle on $\mathbf{CS}^4$ and

$$\mathcal{O}(\overset{k}{\underset{\times}{}} \overset{0}{\underset{\bullet}{\rule{1cm}{0.4pt}}} \overset{0}{\underset{\bullet}{}} \cdots \overset{0}{\underset{\bullet}{}} \overset{0}{\underset{\bullet}{}}) \tag{5}$$

is a bundle on $\mathbf{CP_n}$ (namely the kth tensor power of the hyperplane section bundle, usually denoted by $\mathcal{O}(k)$). Many more examples are given below (and related to standard bundles on various spaces). Before looking at these, note that we shall yet further abuse our long-suffering notation: when Dynkin diagrams are used to indicate irreducible P-modules, the same diagram shall also stand for the sheaf of the associated homogeneous bundle. In other words, where no confusion will arise, we shall omit "$\mathcal{O}(\ )$" in the notation above.

Example (3.2.1). Regarding $\mathbf{g}$ as the left invariant vector fields on G yields an isomorphism

$$\Theta(G/P) \cong \mathcal{O}(\mathbf{g}/\mathbf{p}),$$

identifying the *holomorphic* tangent bundle of G/P as a homogeneous vector bundle. The cases in which this is irreducible were discussed in example 3.1.10 above. They comprise projective spaces, Grassmannians, complex-ified spheres and their twistor spaces (in even dimensions), Lagrangian Grassmannians, and two exceptional examples. Dually, the cotangent bundle is

$$\Omega^1(G/P) \cong \mathcal{O}(\mathbf{g}/\mathbf{p})^* \equiv \mathcal{O}(\mathbf{u}),$$

where $\mathbf{u}$ is the maximal nilpotent part of $\mathbf{p}$, and we have used the Killing form on $\mathbf{g}$ to identify $(\mathbf{g}/\mathbf{p})^* \cong \mathbf{u}$. For both these bundles, table 3.2 may now be interpreted as giving the irreducible cases of $\Theta(G/P)$ and $\Omega^1(G/P)$.

Example (3.2.2). On the complete flag variety $\mathbf{F}_c(V)$ (cf. 2.3.1), we have homogeneous line bundles such as

$$\mathcal{F} = \overset{p}{\times}\!-\!\overset{q}{\times}\!-\!\overset{r}{\times} \cdots \overset{t}{\times}\!-\!\overset{u}{\times}.$$

It is easy to give a simple geometric description of these bundles: at a flag $\{V_i\} \in \mathbf{F}_c(V)$, the stalk of $\mathcal{F}$ is isomorphic to

$$V_1^p \otimes (V_2/V_1)^{p+q} \otimes \ldots \otimes (V_{n-1}/V_{n-2})^{p+q+r+\ldots+t+u}$$

or

$$(V_2/V_1)^p \otimes (V_3/V_2)^{p+q} \otimes \ldots \otimes (V/V_{n-1})^{p+q+r+\ldots+t+u}$$

(since $\det V$ is trivial). A similar identification is possible on partial flag manifolds as in [50]. Thus, for example, on $\bullet\!-\!\times\!-\!\bullet$:

$$\overset{1}{\bullet}\!\overset{0}{-\!\times\!-}\!\overset{1}{\bullet} \cong (0,1)V_x \otimes (1,2)(\mathbf{T}/V_x)$$
$$\cong (-1,0)V_x \otimes (0,1)(\mathbf{T}/V_x)$$
$$\cong V_x^* \otimes \mathbf{T}/V_x,$$

where $V_x \subset \mathbf{T}$ is the two dimensional subspace of $\mathbf{T}$ corresponding to x. This is the tangent bundle to Minkowski space. Another example is

$$\overset{m\ \ 0\ \ 0\qquad\ \ 0\ \ 0}{\times\!\!-\!\!\bullet\!\!-\!\!\bullet\ \cdots\ \bullet\!\!-\!\!\bullet} \cong (0,m)\mathbf{L} \otimes (m,m,\ldots,m)(\mathbf{C}^{n+1}/\mathbf{L})$$
$$\cong (-m,0)\mathbf{L} \otimes (0,0,\ldots,0)(\mathbf{C}^{n+1}/\mathbf{L})$$
$$\cong \mathbf{L}^{\otimes -m}.$$

This is the mth tensor power of the hyperplane section bundle on $\mathbf{CP_n}$, verifying the claim at (5).

Example (3.2.3).   Bundles on Minkowski space
We identify here the standard bundles on Minkowski space $\mathbf{M} = \bullet\!\!-\!\!\times\!\!-\!\!\bullet$ as given, for example, in [44,127]; this provides theoretical physicists with some concrete examples of homogeneous bundles and will serve to introduce others to Penrose's abstract index notation [127].

Spinor bundles

$$\mathcal{O}^{A'} = \overset{1\ \ \ 0\ \ \ 0}{\bullet\!\!-\!\!\times\!\!-\!\!\bullet} \qquad\qquad \mathcal{O}^{A} = \overset{0\ \ \ 0\ \ \ 1}{\bullet\!\!-\!\!\times\!\!-\!\!\bullet}$$

$$\mathcal{O}_{A'} = \overset{1\ \ -1\ \ 0}{\bullet\!\!-\!\!\times\!\!-\!\!\bullet} \qquad\qquad \mathcal{O}_{A} = \overset{0\ \ -1\ \ 1}{\bullet\!\!-\!\!\times\!\!-\!\!\bullet}$$

$$\mathcal{O}[n] = \overset{0\ \ \ n\ \ \ 0}{\bullet\!\!-\!\!\times\!\!-\!\!\bullet} \quad \text{(the conformal weight } n \text{ line bundle)}$$

and, taking tensor products,

$$\mathcal{O}^{\overbrace{(A'B'\ldots D')}^{p}\overbrace{(EF\ldots H)}^{r}}[q] = \overset{p\ \ \ q\ \ \ r}{\bullet\!\!-\!\!\times\!\!-\!\!\bullet} = \mathcal{O}_{\underbrace{(A'B'\ldots D')}_{p}\underbrace{(EF\ldots H)}_{r}}[p+q+r]$$

Tangent bundle on $\mathbf{M}$

$$\Theta(\mathbf{M}) = \mathcal{O}^{AA'} = \overset{1\ \ \ 0\ \ \ 1}{\bullet\!\!-\!\!\times\!\!-\!\!\bullet}$$

Forms on $\mathbf{M}$

$$\Omega^1 \ = \mathcal{O}_{AA'} \qquad\qquad\qquad\quad = \overset{1\ \ -2\ \ 1}{\bullet\!\!-\!\!\times\!\!-\!\!\bullet}$$

$$\Omega^2 \ = \mathcal{O}_{(A'B')}[-1] \oplus \mathcal{O}_{(AB)}[-1] = \overset{2\ \ -3\ \ 0}{\bullet\!\!-\!\!\times\!\!-\!\!\bullet} \oplus \overset{0\ \ -3\ \ 2}{\bullet\!\!-\!\!\times\!\!-\!\!\bullet}$$

$$\Omega^3 \ = \mathcal{O}_{AA'}[-2] \qquad\qquad\quad\ = \overset{1\ \ -4\ \ 1}{\bullet\!\!-\!\!\times\!\!-\!\!\bullet}$$

$$\Omega^4 \ = \mathcal{O}[-4] \qquad\qquad\qquad\ = \overset{0\ \ -4\ \ 0}{\bullet\!\!-\!\!\times\!\!-\!\!\bullet}$$

$$\odot^2\Omega^1 = \mathcal{O}_{(AB)(A'B')} \oplus \mathcal{O}[-2] \quad = \overset{2\ \ -4\ \ 2}{\bullet\!\!-\!\!\times\!\!-\!\!\bullet} \oplus \overset{0\ \ -2\ \ 0}{\bullet\!\!-\!\!\times\!\!-\!\!\bullet}$$

Example (3.2.4).   Bundles on $\mathbf{CS}^{2n}$ and $\mathbf{CS}^{2n+1}$

Generalizing from above, it is natural to make the following definitions on $\mathbf{CS}^{2n}$ and $\mathbf{CS}^{2n+1}$.

Conformal weights on $\mathbf{CS}^{2n}$:

$$\mathcal{O}[q] = \;\overset{q}{\underset{\times}{}} \overset{0}{\bullet} \overset{0}{\bullet} \cdots \overset{0}{\bullet} \overset{0}{\bullet} \overset{0}{\diagup} \underset{0}{\diagdown}$$

Sections of this line bundle may be represented locally, given a choice of metric g, by functions f which rescale according to $f \mapsto \kappa^q f$ when g is replaced by $\kappa^2 g$. g must be a metric in the *conformal* class of metrics—see remark 3.2.5 below.

Spinor bundles on $\mathbf{CS}^{2n}$:

$$\mathcal{O}^{\alpha'} = \;\overset{0}{\underset{\times}{}} \overset{0}{\bullet} \overset{0}{\bullet} \cdots \overset{0}{\bullet} \overset{0}{\bullet} \overset{1}{\diagup} \underset{0}{\diagdown}$$

$$\mathcal{O}^{\alpha} = \;\overset{0}{\underset{\times}{}} \overset{0}{\bullet} \overset{0}{\bullet} \cdots \overset{0}{\bullet} \overset{0}{\bullet} \overset{0}{\diagup} \underset{1}{\diagdown}$$

If n is *odd*, these are dual, up to a conformal factor:

$$\mathcal{O}_{\alpha'} = \;\overset{-1}{\underset{\times}{}} \overset{0}{\bullet} \overset{0}{\bullet} \cdots \overset{0}{\bullet} \overset{0}{\bullet} \overset{0}{\diagup} \underset{1}{\diagdown}$$

$$\mathcal{O}_{\alpha} = \;\overset{-1}{\underset{\times}{}} \overset{0}{\bullet} \overset{0}{\bullet} \cdots \overset{0}{\bullet} \overset{0}{\bullet} \overset{1}{\diagup} \underset{0}{\diagdown}$$

If n is *even*, the spinor bundles are self-dual, again up to a conformal factor:

$$\mathcal{O}_{\alpha'} = \;\overset{-1}{\underset{\times}{}} \overset{0}{\bullet} \overset{0}{\bullet} \cdots \overset{0}{\bullet} \overset{0}{\bullet} \overset{1}{\diagup} \underset{0}{\diagdown}$$

$$\mathcal{O}^{\alpha} = \;\overset{-1}{\underset{\times}{}} \overset{0}{\bullet} \overset{0}{\bullet} \cdots \overset{0}{\bullet} \overset{0}{\bullet} \overset{0}{\diagup} \underset{1}{\diagdown}$$

On $\mathbf{CS^{2n+1}}$ there is only one type of spinor:

$$\mathcal{O}^\alpha = \overset{0}{\underset{\times}{}}\!\!-\!\!\overset{0}{\bullet}\!\!-\!\!\overset{0}{\bullet}\ \cdots\ \overset{0}{\bullet}\!\!\Rightarrow\!\!\overset{1}{\bullet}$$

with dual

$$\mathcal{O}_\alpha = \overset{-1}{\underset{\times}{}}\!\!-\!\!\overset{0}{\bullet}\!\!-\!\!\overset{0}{\bullet}\ \cdots\ \overset{0}{\bullet}\!\!\Rightarrow\!\!\overset{1}{\bullet}\ .$$

The tangent bundles are given in table 3.2; it is useful to work out the bundles of p-forms. A simple method of doing this is given in chapter 8, but in any event it is not too difficult to compute them directly by taking exterior powers of the one-forms.

   Forms on $\mathbf{CS^{2n}}$:

$$\Omega^p = \overset{-p\text{-}1}{\underset{\times}{}}\ \overset{0}{\bullet}\ \overset{0}{\bullet}\ \cdots\ \overset{0}{\bullet}\ \overset{1}{\bullet}\ \overset{0}{\bullet}\ \cdots\ \overset{0}{\bullet}\ \overset{0}{\bullet}\!\!<\!\!\begin{smallmatrix}0\\[4pt]0\end{smallmatrix}\qquad 0<p<n-1$$
$$\text{node } p+1 \uparrow$$

$$\Omega^{n-1} = \overset{-n}{\underset{\times}{}}\ \overset{0}{\bullet}\ \overset{0}{\bullet}\ \cdots\ \overset{0}{\bullet}\ \overset{0}{\bullet}\!\!<\!\!\begin{smallmatrix}1\\[4pt]1\end{smallmatrix}$$

$$\Omega^{n} = \overset{-n\text{-}1}{\underset{\times}{}}\ \overset{0}{\bullet}\ \overset{0}{\bullet}\ \cdots\ \overset{0}{\bullet}\ \overset{0}{\bullet}\!\!<\!\!\begin{smallmatrix}2\\[4pt]0\end{smallmatrix}\ \oplus\ \overset{-n\text{-}1}{\underset{\times}{}}\ \overset{0}{\bullet}\ \overset{0}{\bullet}\ \cdots\ \overset{0}{\bullet}\ \overset{0}{\bullet}\!\!<\!\!\begin{smallmatrix}0\\[4pt]2\end{smallmatrix}\ = \Omega^n_+ \oplus \Omega^n_-$$

$$\Omega^{n+1} = \overset{-n\text{-}2}{\underset{\times}{}}\ \overset{0}{\bullet}\ \overset{0}{\bullet}\ \cdots\ \overset{0}{\bullet}\ \overset{0}{\bullet}\!\!<\!\!\begin{smallmatrix}1\\[4pt]1\end{smallmatrix}$$

$$\Omega^p = \overset{-p\text{-}1}{\underset{\times}{}}\ \overset{0}{\bullet}\ \overset{0}{\bullet}\ \cdots\ \overset{0}{\bullet}\ \overset{1}{\bullet}\ \overset{0}{\bullet}\ \cdots\ \overset{0}{\bullet}\ \overset{0}{\bullet}\!\!<\!\!\begin{smallmatrix}0\\[4pt]0\end{smallmatrix}\qquad n+1<p<2n$$
$$\text{node } 2n+1-p \uparrow$$

$$\Omega^{2n} = \overset{-2n}{\underset{\times}{}}\ \overset{0}{\bullet}\ \overset{0}{\bullet}\ \cdots\ \overset{0}{\bullet}\ \overset{0}{\bullet}\!\!<\!\!\begin{smallmatrix}0\\[4pt]0\end{smallmatrix}$$

Remark (3.2.5).   Regarded as generalized flag manifolds, the complex spheres $\mathbf{CS}^{2n}$ come with a conformal structure; this is the distinguished line bundle

$$
\begin{array}{ccccccc}
-2 & 0 & 0 & & 0 & 0 & \bullet\,0 \\
\times\!\!-\!\!\bullet\!\!-\!\!\bullet & \cdots & \bullet\!\!-\!\!\bullet\!\!\raisebox{2pt}{\diagdown} & \\
& & & & & & \bullet\,0
\end{array}
$$

in $\odot^2\Omega^1$, the bundle whose sections are (complex) metrics. A choice of a metric g in the conformal class together with a choice of complex orientation gives a *Hodge operator*

$$
* \; : \; \Omega^p \overset{\cong}{\to} \Omega^{2n-p}. \tag{6}
$$

This scales by

$$
* \mapsto \kappa^{2(p-n)} * \quad \text{when } g \mapsto \kappa^2 g.
$$

The isomorphism (6) together with this scaling is evident in the above enumeration of the forms as homogeneous bundles. When $p = n$, $*$ depends only on the conformal structure. $*^2 = \pm 1$ and so $*$ has two eigenvalues. $\Omega^n = \Omega^n_+ \oplus \Omega^n_-$ is a splitting into the eigenspaces of $*$, i.e., the self- and anti-self-dual n-forms.

Forms on $\mathbf{CS}^{2n+1}$:

$$
\Omega^p \;=\;
\begin{array}{cccccccccc}
-p\text{-}1 & 0 & 0 & & 0 & 1 & 0 & & 0 & 0 \\
\times\!\!-\!\!\bullet\!\!-\!\!\bullet & \cdots & \bullet\!\!-\!\!\bullet\!\!-\!\!\bullet & \cdots & \bullet\!\!\Rrightarrow\!\!\bullet
\end{array}
\quad 0 < p < n
$$

node $p+1 \uparrow$

$$
\Omega^n \;=\;
\begin{array}{cccccc}
-n\text{-}1 & 0 & 0 & & 0 & 2 \\
\times\!\!-\!\!\bullet\!\!-\!\!\bullet & \cdots & \bullet\!\!\Rrightarrow\!\!\bullet
\end{array}
$$

$$
\Omega^{n+1} \;=\;
\begin{array}{cccccc}
-n\text{-}2 & 0 & 0 & & 0 & 2 \\
\times\!\!-\!\!\bullet\!\!-\!\!\bullet & \cdots & \bullet\!\!\Rrightarrow\!\!\bullet
\end{array}
$$

$$
\Omega^p \;=\;
\begin{array}{cccccccccc}
-p\text{-}1 & 0 & 0 & & 0 & 1 & 0 & & 0 & 0 \\
\times\!\!-\!\!\bullet\!\!-\!\!\bullet & \cdots & \bullet\!\!-\!\!\bullet\!\!-\!\!\bullet & \cdots & \bullet\!\!\Rrightarrow\!\!\bullet
\end{array}
\quad n+1 < p < 2n+1
$$

node $p+1 \uparrow$

$$
\Omega^{2n+1} \;=\;
\begin{array}{cccccc}
-2n\text{-}1 & 0 & 0 & & 0 & 0 \\
\times\!\!-\!\!\bullet\!\!-\!\!\bullet & \cdots & \bullet\!\!\Rrightarrow\!\!\bullet
\end{array}
$$

3.3   A remark on inverse images

As in section 2.4, suppose that $Q \subset P$ so that there is a fibration

$$
G/Q \overset{\eta}{\to} G/P.
$$

Then there is a natural map of sheaves

$$0 \to \eta^{-1}\mathcal{O}_{\mathbf{p}}(\lambda) \to \mathcal{O}_{\mathbf{q}}(\lambda) \tag{7}$$

induced by the projection $E_{\mathbf{p}}(\lambda) \to E_{\mathbf{q}}(\lambda)$. Irreducibility of the representation $E_{\mathbf{p}}(\lambda)$ yields injectivity of the map. (7) is the beginning of a resolution of $\eta^{-1}\mathcal{O}_{\mathbf{p}}(\lambda)$ on G/Q by homogeneous vector bundles, which will be completed in chapter 8.

THE WEYL GROUP, ITS ACTIONS, AND HASSE DIAGRAMS

The combinatorial key to the *Penrose transform* is the *Weyl group* of a semisimple Lie algebra and the *Hasse diagram* or subgraph associated to a parabolic subalgebra, together with their action on weights. With this machinery, we may describe the two key ingredients of the transform—the *Bernstein–Gelfand–Gelfand resolutions* and the *Bott–Borel–Weil* theorem for computing cohomology and direct images of homogeneous sheaves. Again, the forbearance of the reader who is familiar with this material is requested whilst we review it for the mathematical physicist. A good reference for what follows is the excellent book by Humphreys [94].

4.1 The Weyl group

Recall that on $\mathbf{h}_{\mathbf{R}}^*$, the bilinear form $\langle\,,\,\rangle$ is positive definite. For any $\alpha \in \Delta(\mathbf{g})$ the *wall* W_α, as in section 3.1, is the hyperplane perpendicular to α. Let σ_α be the reflection in W_α : then the *Weyl group* of $\mathbf{g}$, $W_{\mathbf{g}}$ or simply W if $\mathbf{g}$ is understood by context, is the group generated by the σ_α for $\alpha \in \Delta(\mathbf{g})$. In fact it is sufficient to deal with simple roots only—if S is a fixed system of such for $\mathbf{g}$, then $\{\sigma_\alpha$ s.t. $\alpha \in S\}$ generates $W_{\mathbf{g}}$. Given any $w \in W$, there exists a minimal integer $\ell(w)$, called the *length* of w, such that w can be expressed as a product of $\ell(w)$ simple reflections. Such an expression, which is generally not unique, is called a *reduced expression* for w. W is finite, and there is an unique element w^0 of maximal length.

Example (4.1.1).   [The Weyl group of $\mathbf{sl}(3, \mathbf{C})$] Let $\mathbf{g} = \mathbf{sl}(3, \mathbf{C})$ with simple roots numbered as $\overset{\alpha_1}{\bullet}\!\!-\!\!\overset{\alpha_2}{\bullet}$, and denote by $(i \ldots k)$ the product

$\sigma_{\alpha_i} \ldots \sigma_{\alpha_k}$. Then the reader may verify that reduced expressions for elements of $W_{\mathbf{g}}$ are

$$\{\mathrm{id}, (1), (2), (12), (21), (121)\}.$$

Notice that $(121) = (212)$. The reader should check that $W_{\mathbf{g}} = S_3$, the symmetric group on three letters; σ_{α_1} and σ_{α_2} correspond to the simple transpositions $\{12\}$ and $\{23\}$. Generally, if $\mathbf{g} = \mathbf{sl}(n, \mathbf{C})$, then $W_{\mathbf{g}} = S_{n+1}$.

Remark (4.1.2).   Simple realizations of Weyl groups for various $\mathbf{g}$ may be found in several places—for example [85]. $W_{\mathbf{g}}$ is a particular example of a class of finite groups called *Coxeter groups*.

W acts on $\mathbf{h}_{\mathbf{R}}^*$, by definition. It permutes the set Δ of roots of $\mathbf{g}$. The orbit of any root consists of roots of the same length, since reflections preserve $\langle\ ,\ \rangle$ and, in fact, consists of *all* such roots. As in section 3.1, the walls W_α for $\alpha \in \Delta$ carve up $\mathbf{h}_{\mathbf{R}}^*$ into cones known as *Weyl chambers*, all of which are congruent. The Weyl group acts faithfully on these chambers. In other words there is precisely one element of W which maps any given chamber into any other. Thus, if one chamber is arbitrarily chosen then there is a one-to-one correspondence between these chambers and the elements of W in which the identity element of W corresponds to the chosen chamber. In particular, we can take this to be the fundamental or dominant chamber as in section 3.1. Recall that the dominant chamber is obtained from a choice of simple roots. Conversely, any chamber determines a set of simple roots. The action of the Weyl group on the chambers corresponds to its action on choices of simple root systems.

W also permutes the weights in any finite dimensional irreducible representation of $\mathbf{g}$. The highest weight lies in the closure of the dominant chamber, the lowest weight is obtained by action of the longest element w^0 of W, and the orbit of the highest weight under W is called the set of *extremal* weights. Thus, an extremal weight λ determines the representation, which, as in remark 3.1.5, we denote by $F(\lambda)$.

To realize this reflection action of W, we shall need an explicit form for the action of a reflection σ_α on a weight λ, written in the Dynkin diagram notation. If $\lambda \in \mathbf{h}_{\mathbf{R}}^*$, then:

$$\sigma_\alpha(\lambda) = \lambda - \langle \lambda, \alpha^\vee \rangle \alpha$$

(recall that $\alpha^\vee = 2\alpha/\langle \alpha, \alpha \rangle$ is the *co-root* of α). Thus, the node coefficients for $\sigma_\alpha(\lambda)$ are given by the formula

$$\langle \sigma_\alpha(\lambda), \alpha_i^\vee \rangle = \langle \lambda, \alpha_i^\vee \rangle - \langle \lambda, \alpha^\vee \rangle \langle \alpha, \alpha_i^\vee \rangle$$

where the α_i range over S. If α is simple, $\langle \alpha, \alpha_i^\vee \rangle$ is a *Cartan integer* obtainable directly from the Dynkin diagram as in section 2.1. This yields the following:

Recipe (4.1.3).   [The action of a simple reflection]   *To compute $\sigma_\alpha(\lambda)$, let c be the coefficient of the node associated to α. Add c to the adjacent coefficients, with multiplicity if there is a multiple edge directed towards the adjacent node, and then replace c by $-c$.*

Example (4.1.4).   Reflect in the simple wall W_α indicated by ↑:

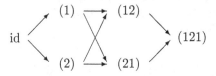

The Weyl group $W_{\mathbf{g}}$ admits the structure of a *directed graph* as follows: let $w, w' \in W_{\mathbf{g}}$ and write

$$w \to w'$$

if $\ell(w') = \ell(w) + 1$ and $w' = \sigma_\alpha w$ for some $\alpha \in \Delta(\mathbf{g})$. These are precisely the directed edges in $W_{\mathbf{g}}$. This directed graph structure gives a partial ordering known as the *Bruhat* ordering [18] on $W_{\mathbf{g}}$: we write $w \preceq w'$ if either $w = w'$ or there exists a directed path from w to w' in $W_{\mathbf{g}}$.

Example (4.1.5).   [Directed graph structure of $W_{\mathrm{sl}(3,\mathbf{C})}$]

$$
\begin{array}{ccc}
 & (1) \longrightarrow (12) & \\
\mathrm{id} \Big\langle & \times & \Big\rangle (121) \\
 & (2) \longrightarrow (21) &
\end{array}
$$

This directed graph structure is important for several reasons. As will be seen below, it is closely related to the topology of a generalized flag manifold, by specifying a stratification by affine cells of increasing complex dimension. It also restricts the possibility of the existence of invariant differential operators between homogeneous sheaves, for these correspond to homomorphisms of *Verma modules* (see section 11.1 and [17,39,53,110,150]), which are restricted to lie between Verma modules whose highest weights are related by the *affine action* of $W_{\mathbf{g}}$ given below. The differential operators which occur in the *Penrose transform* are precisely of this form (and this is one reason for studying it). It is therefore important to have at hand a method for computing the structure.

The actual computation of $W_{\mathbf{g}}$ from its definition is extremely tedious. A straightforward alternative (which will be generalized in the next section) is obtained by noting that the elements of the orbit of any *regular* weight are in one-to-one correspondence with $W_{\mathbf{g}}$. This follows because a regular weight is defined to be one which lies on no wall. In particular, set $\rho = \frac{1}{2}\sum_{\alpha\in\Delta^+(\mathbf{g})}\alpha$. It is readily checked that $\langle\rho,\alpha_i^\vee\rangle = 1$ for all simple α_i so that $\rho = \sum_i \lambda_i$. Thus the Dynkin diagram for ρ has a 1 over each node. Geometrically, ρ is the integral weight in the dominant chamber which lies closest to the origin, i.e., right in the corner of the chamber. We may now compute the orbit of ρ by repeatedly applying simple reflections (using the *recipe* given above). This results in a one-to-one correspondence with the elements of $W_{\mathbf{g}}$. Reduced expressions for such an element are then obtained by tracing back all possible concatenations of simple reflections producing an element of the orbit from ρ.

Example (4.1.6).   [Computing the Weyl group of $\mathbf{sl}(4,\mathbf{C})$]
The diagram in table 4.1 shows the action of $W_{\mathbf{sl}(4,\mathbf{C})}$ on $\rho = \overset{1}{\bullet}\!\!-\!\!\overset{1}{\bullet}\!\!-\!\!\overset{1}{\bullet}$ where the lines denote the application of a simple reflection according to recipe 4.1.3. Other reflections are not recorded on this diagram (see example 4.1.9 below).

To obtain the directed graph structure of $W_{\mathbf{g}}$, we make use of the following lemma [18]:

Lemma (4.1.7).   *If $w \to w'$, then a reduced expression for w is obtained from any reduced expression for w' by omitting one simple reflection. More generally, if $w \preceq w'$, then a reduced expression for w is obtained by omitting several simple reflections.*

Remark (4.1.8).   Notice that simply omitting a reflection from a reduced expression for w' may produce w with $\ell(w) < \ell(w') - 1$ so that $w \not\to w'$—consider omitting the (2) from (121) in example 4.1.1 above, for example.

Since the method of computing $W_{\mathbf{g}}$ yields all possible reduced expressions for elements of $W_{\mathbf{g}}$, it is feasible, with care, to apply this lemma. Let us do this for the Weyl group of $\mathbf{sl}(4,\mathbf{C})$:

Table 4.1. Orbit of ρ under $W_{\mathbf{sl}(4,\mathbf{C})}$

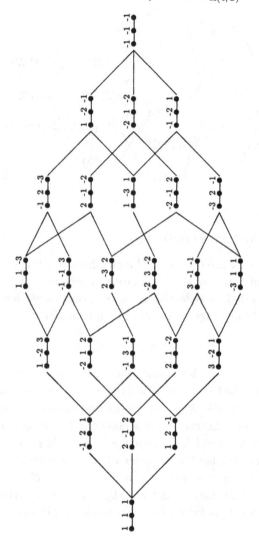

Example (4.1.9).   [Directed graph structure of $W_{\mathrm{sl}(4,\mathbb{C})}$]

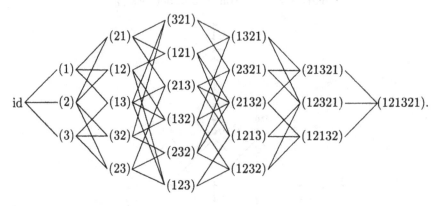

4.2   The affine Weyl action

It turns out, as a consequence of Harish Chandra's theorem [94], that the most significant action of the Weyl group on weights is not the straightforward one generated by reflection as in the previous section but rather the *affine* action obtained by conjugation with translation by $\rho = \sum \lambda_i$:

$$w.\lambda = w(\lambda + \rho) - \rho.$$

It is this action which determines the correct parameters in calculating cohomology (via the Bott–Borel–Weil theorem) or resolutions (of Bernstein, Gelfand, and Gelfand). A crucial distinction occurs in both these results which depends on whether λ has a non-trivial stabilizer under the affine action: if it does, we call λ *singular* and $\lambda + \rho$ is in some wall. Otherwise, λ is *non-singular* (rather than regular which means that λ itself is not in a wall). If λ is singular, then for some $w \in W$ and $\alpha_i \in \mathcal{S}$, $\langle w(\lambda + \rho), \alpha_i^\vee \rangle = 0$. So a singular weight may be detected by applying the reflection recipe to the Dynkin diagram notation for $\lambda + \rho$ and seeing if a node integer can be made zero.

4.3   The Hasse diagram of a parabolic subalgebra

The Weyl group introduced above has an important subgraph or *Hasse diagram* W^p (see, for example, [35]) attached to a standard parabolic subalgebra **p** of **g**. This diagram is related to the topological structure of the generalized flag manifold G/P and the structure of the *Bernstein–Gelfand–Gelfand* resolution on G/P explained in chapter 8. It is important also in determining the direct images of homogeneous sheaves under projections between generalized flag manifolds (using the *Bott–Borel–Weil* theorem as in chapter 5).

It is distinguished as the subset of $W_{\mathbf{g}}$ whose action sends a weight λ, dominant for $\mathbf{g}$, to a weight dominant for $\mathbf{p}$. To elucidate this condition, let $\mathbf{p} = \mathbf{l} \oplus \mathbf{u}$ be the Levi decomposition of $\mathbf{p}$ given by the fixed choice of the Cartan subalgebra $\mathbf{h}$ and the system of simple roots $S = \{\alpha_i\}$. Let $\{\lambda_i\}$ be the basis of weights dual to $\{\alpha_i^\vee\}$ under $\langle\,,\,\rangle$, as before. Then if $w \in W^{\mathbf{p}}$ and if λ is dominant for $\mathbf{g}$, we must have

$$\langle \lambda, w^{-1}\alpha \rangle = \langle w\lambda, \alpha \rangle \geq 0 \ \ \forall \alpha \in \Delta^+(\mathbf{l}).$$

Letting λ run over $\{\lambda_i\}$ forces $w^{-1}\alpha \in \Delta^+(\mathbf{g})$. This is clearly sufficient, also, for $w\lambda$ to be **p**-dominant and hence for $w \in W^{\mathbf{p}}$. Defining

$$\Delta(w) = \{\alpha \in \Delta^+(\mathbf{g}) \text{ s.t. } w^{-1}\alpha \in -\Delta^+(\mathbf{g})\},$$

we obtain

$$W^{\mathbf{p}} = \{w \in W_{\mathbf{g}} \text{ s.t. } \Delta(w) \subseteq \Delta(\mathbf{u})\}.$$

Remark (4.3.1).

1. Note that $|\Delta(w)| = \ell(w)$ as in [94, pp. 51–52].

2. $\Delta(w)$ and its complement $\Delta^+(\mathbf{g})$ both have the property that a sum of two elements in the set is either in the set or is not a root. $\Delta(\cdot)$ is a bijection of $W_{\mathbf{g}}$ with such sets [102]. Complementation then corresponds to multiplication by w^0, the longest element of $W_{\mathbf{g}}$.

3. We induce the *subgraph* structure on $W^{\mathbf{p}}$.

Example (4.3.2).   [Minkowski space] If $\mathbf{p} = \bullet\!\!-\!\!\times\!\!-\!\!\bullet$, then, as can be seen directly from examples 4.1.6 and 4.1.9, $W^{\mathbf{p}}$ is

$$
\text{id} \longrightarrow (2)
\begin{array}{c}
\nearrow (21) \searrow \\
\\
\searrow (23) \nearrow
\end{array}
(213) \longrightarrow (2132).
$$

Computing $W^{\mathbf{p}}$ from the definition is usually lengthy. However, the method introduced above to compute the Weyl group $W_{\mathbf{g}}$ is readily modified to yield a method which can be very efficient indeed. To give it, we need an alternative characterization of $W^{\mathbf{P}}$, due to Kostant [102]. This identifies $W^{\mathbf{P}}$ with the set of minimal length right coset representatives of the subgroup $W_{\mathbf{p}}$ of $W_{\mathbf{g}}$ generated by simple reflections σ_α for $\alpha \in \mathcal{S}_{\mathbf{p}}$ ($W_{\mathbf{p}}$ is just the Weyl group of a reductive Levi factor of $\mathbf{p}$).

Lemma (4.3.3).   [Kostant] *Any $w \in W_{\mathbf{g}}$ admits a unique decomposition $w = w_{\mathbf{p}} w^{\mathbf{P}}$ with $w_{\mathbf{p}} \in W_{\mathbf{p}}$ and $w^{\mathbf{P}} \in W^{\mathbf{P}}$. Moreover, $\ell(w) = \ell(w_{\mathbf{p}}) + \ell(w^{\mathbf{P}})$ and $w^{\mathbf{P}}$ is the only element of the right coset $W_{\mathbf{p}} w^{\mathbf{P}}$ with this minimal length.*

To use this, let $\rho^{\mathbf{P}} = \sum \lambda_i$, where i ranges such that $\alpha_i \in \mathcal{S} \setminus \mathcal{S}_{\mathbf{p}}$. Thus, the Dynkin diagram for $\rho^{\mathbf{P}}$ has ones over nodes crossed through in the diagram for $\mathbf{p}$ and zeros elsewhere. The idea is to identify the orbit of $\rho^{\mathbf{P}}$ under $W_{\mathbf{g}}$ with $W^{\mathbf{P}}$ just as the orbit of ρ is identified with $W_{\mathbf{g}}$.

Lemma (4.3.4).   *The stabilizer of $\rho^{\mathbf{P}}$ in $W_{\mathbf{g}}$ is $W_{\mathbf{p}}$.*

Proof   Certainly $W_{\mathbf{p}}$ stabilizes $\rho^{\mathbf{P}}$, so we need to prove that $w\rho^{\mathbf{P}} = \rho^{\mathbf{P}} \Rightarrow w \in W_{\mathbf{p}}$. By Kostant's lemma (4.3.3), we can write $w = w_{\mathbf{p}} w^{\mathbf{P}}$, in which case $w^{\mathbf{P}} \rho^{\mathbf{P}} = \rho^{\mathbf{P}}$ and it suffices to shew that then $w^{\mathbf{P}}$ is the identity. By definition, $w^{\mathbf{P}}$ takes any weight dominant for $\mathbf{g}$ to a weight dominant for $\mathbf{p}$. In particular, $w^{\mathbf{P}} \rho$ is dominant for $\mathbf{p}$—in terms of Dynkin diagrams, $w^{\mathbf{P}} \rho$ has non-negative integers over the uncrossed nodes. Hence, for sufficiently large N, $w^{\mathbf{P}} \rho + N \rho^{\mathbf{P}}$ is dominant for $\mathbf{g}$. However, this is $w^{\mathbf{P}}(\rho + N\rho^{\mathbf{P}})$ and $\rho + N\rho^{\mathbf{P}}$ is strictly dominant for $\mathbf{g}$. Since $W_{\mathbf{g}}$ acts faithfully on the Weyl chambers, this forces $w^{\mathbf{P}}$ to be the identity.   □

Thus, the orbit of $\rho^{\mathbf{P}}$ under the *right* action of $W_{\mathbf{g}}$ given by

$$(\rho^{\mathbf{P}}, w) \to w^{-1} \rho^{\mathbf{P}}$$

is in one-to-one correspondence with $W^{\mathbf{P}}$, by Kostant's lemma. To trace out the orbit, we again repeatedly apply simple reflections to $\rho^{\mathbf{P}}$. Every member of the orbit is connected to $\rho^{\mathbf{P}}$ by one or more paths of such simple reflections. The corresponding member of $W^{\mathbf{P}}$ has reduced expressions which are obtained by taking their composition *in the reverse order* (to account for the inverse in the above action). This is illustrated in the following examples.

Example (4.3.5).   [Minkowski space] If $\mathbf{p} = \ \bullet\!\!-\!\!\times\!\!-\!\!\bullet\ $ as in example 4.3.2, then the orbit of $\rho^{\mathbf{P}}$ under the Weyl group of $\mathbf{sl}(4, \mathbf{C})$ is

$$
\begin{array}{c}
0\ 1\ 0 \\ \bullet\!-\!\times\!-\!\bullet
\end{array}
\longrightarrow
\begin{array}{c}
1\ \text{-}1\ 1 \\ \bullet\!-\!\times\!-\!\bullet
\end{array}
\Big\langle
\begin{array}{c}
\text{-}1\ \ 0\ \ 1 \\ \bullet\!-\!\times\!-\!\bullet \\[4pt]
1\ \ 0\ \ \text{-}1 \\ \bullet\!-\!\times\!-\!\bullet
\end{array}
\Big\rangle
\begin{array}{c}
\text{-}1\ 1\ \text{-}1 \\ \bullet\!-\!\times\!-\!\bullet
\end{array}
\longrightarrow
\begin{array}{c}
0\ \text{-}1\ 0 \\ \bullet\!-\!\times\!-\!\bullet
\end{array}.
$$

This gives $W^\mathbf{P}$ as in example 4.3.2. For later use (in section 8.5.1) notice that the *left* action of the elements of $W^\mathbf{P}$ on the dominant weight $\rho = \begin{smallmatrix} 1\ \ 1\ \ 1 \\ \bullet\!-\!\times\!-\!\bullet \end{smallmatrix}$ is

$$
\begin{array}{c}
1\ 1\ 1 \\ \bullet\!-\!\times\!-\!\bullet
\end{array}
\longrightarrow
\begin{array}{c}
2\ \text{-}1\ 2 \\ \bullet\!-\!\times\!-\!\bullet
\end{array}
\Big\langle
\begin{array}{c}
1\ \ \text{-}2\ \ 3 \\ \bullet\!-\!\times\!-\!\bullet \\[4pt]
3\ \ \text{-}2\ \ 1 \\ \bullet\!-\!\times\!-\!\bullet
\end{array}
\Big\rangle
\begin{array}{c}
2\ \text{-}3\ 2 \\ \bullet\!-\!\times\!-\!\bullet
\end{array}
\longrightarrow
\begin{array}{c}
1\ \text{-}3\ 1 \\ \bullet\!-\!\times\!-\!\bullet
\end{array}.
$$

Notice that these weights are all dominant for $\mathbf{p}$ as in our original definition of $W^\mathbf{P}$.

Example (4.3.6).   [Complex projective space]

Consider $\mathbf{p} = \begin{smallmatrix} \times\!-\!\bullet\!-\!\bullet \ \cdots\ \bullet\!-\!\bullet \end{smallmatrix}$; $\rho^\mathbf{P} = \begin{smallmatrix} 1\ \ 0\ \ 0 \ \ \ \ 0\ \ 0 \\ \times\!-\!\bullet\!-\!\bullet\ \cdots\ \bullet\!-\!\bullet \end{smallmatrix}$ (it is convenient to carry the cross rather than just use $\begin{smallmatrix} 1\ \ 0\ \ 0\ \ \ \ 0\ \ 0 \\ \bullet\!-\!\bullet\!-\!\bullet\ \cdots\ \bullet\!-\!\bullet \end{smallmatrix}$). We compute the orbit of $\rho^\mathbf{P}$ under $W_\mathbf{g}$ (using the recipe) as:

$$
\begin{array}{c}
1\ \ 0\ \ 0\ \ \ \ 0\ \ 0 \\ \times\!-\!\bullet\!-\!\bullet\ \cdots\ \bullet\!-\!\bullet
\end{array}
\to
\begin{array}{c}
\text{-}1\ \ 1\ \ 0\ \ \ \ 0\ \ 0 \\ \times\!-\!\bullet\!-\!\bullet\ \cdots\ \bullet\!-\!\bullet
\end{array}
\to
\begin{array}{c}
0\ \ \text{-}1\ \ 1\ \ \ \ 0\ \ 0 \\ \times\!-\!\bullet\!-\!\bullet\ \cdots\ \bullet\!-\!\bullet
\end{array}
\to
$$

$$
\cdots
\begin{array}{c}
0\ \ 0\ \ 0\ \ \ \ \text{-}1\ \ 1 \\ \times\!-\!\bullet\!-\!\bullet\ \cdots\ \bullet\!-\!\bullet
\end{array}
\to
\begin{array}{c}
0\ \ 0\ \ 0\ \ \ \ 0\ \ \text{-}1 \\ \times\!-\!\bullet\!-\!\bullet\ \cdots\ \bullet\!-\!\bullet
\end{array}
$$

from which the Hasse diagram for $\mathbf{p}$ follows as

$$
\mathrm{id} \to (1) \to (12) \to (123) \to \cdots \to (123\ldots n).
$$

Example (4.3.7).   [Even dimensional spheres]

Let $\mathbf{p} =$ ⌧————•——• ··· •—< with $n+1$ nodes numbered from left to

right and top to bottom. Then $W^{\mathbf{p}}$ has a single element of each length $0, 1, \ldots, n-1, n+1, \ldots 2n$ and two elements of length n as follows:

element	length
$s_0 = id$	0
$s_j = \sigma_1\sigma_2\ldots\sigma_j$	$j, \quad 1 \le j \le n-1$
$s_+ = \sigma_1\sigma_2\ldots\sigma_{n-1}\sigma_n$	n
$s_- = \sigma_1\sigma_2\ldots\sigma_{n-1}\sigma_{n+1}$	n
$s_{n+1} = \sigma_1\sigma_2\ldots\sigma_{n-1}\sigma_n\sigma_{n+1}$	$n+1$
$s_{n+j} = \sigma_1\sigma_2\ldots\sigma_{n-1}\sigma_n\sigma_{n+1}\sigma_{n-1}\sigma_{n-2}\ldots\sigma_{n+1-j}$	$n+j, \quad 2 \le j \le n$

giving rise to the following picture:

for $W^{\mathbf{p}}$ as a graph. Notice, from example 3.2.4, that the de Rham sequence on $\mathbf{CS^{2n}}$ displays exactly the same pattern. This is not a mere coincidence (see chapter 8).

Example (4.3.8).   [Odd dimensional spheres]
Let $\mathbf{p} =$ ⌧————•——• ··· •⇒• again with $n+1$ nodes numbered from left to right. Then $W^{\mathbf{p}}$ has a single element of each length from zero to $2n+1$:

element	length
$r_0 = id$	0
$r_j = \sigma_1\sigma_2\ldots\sigma_j$	$j, \quad 1 \le j \le n+1$
$r_{n+j} = \sigma_1\sigma_2\ldots\sigma_n\sigma_{n+1}\sigma_n\sigma_{n-1}\ldots\sigma_{n+2-j}$	$n+j, \quad 2 \le j \le n+1$.

Remark (4.3.9).   The weights for $\mathbf{g}$ traced out in computing $W^{\mathbf{p}}$ are the extremal weights of a finite dimensional irreducible $\mathbf{g}$-module, of highest weight $\rho^{\mathbf{p}}$.

Remark (4.3.10).   The cardinality $|W^{\mathbf{p}}| = |W_{\mathbf{g}}|/|W_{\mathbf{p}}|$, by Kostant's lemma. The table on page 66 of [94] gives $|W_{\mathbf{g}}|$ for $\mathbf{g}$ simple. Of course, $\dim G/P = |\Delta^+(\mathbf{g})| - |\Delta^+(\mathbf{l_s})|$, and these cardinalities are also given in the table. This difference is also the length of the unique longest element $w_{\mathbf{p}}^0$ in $W^{\mathbf{p}}$ since $\Delta(w_{\mathbf{p}}^0) = \Delta(\mathbf{u})$.

4.4   Relative Hasse diagrams

In the sequel, we shall usually need to apply Bernstein–Gelfand–Gelfand resolutions and the Bott–Borel–Weil theorem to the fibres of a double fibration; we have already seen that these are generalized flag varieties in their own right. To do this we shall need to work with a relative Hasse diagram which we shall now define. As in section 2.4, consider the fibration

$$ G/Q \xrightarrow{\ \tau\ } G/P $$

and recall that the τ fibres have the form $P/Q = L_S/L_S \cap Q$ where L_S is the semisimple part of a reductive Levi factor of P. The Weyl group of the Lie algebra of L_S is $W_{\mathbf{p}}$ and this plays the rôle of $W_{\mathbf{g}}$ in fibrewise computations. The only difference is that $W_{\mathbf{p}}$, as a subgroup of $W_{\mathbf{g}}$, acts on weights for $\mathbf{g}$. This is important in later calculations where, in terms of our Dynkin diagram notation, we must keep track of what happens to coefficients over the crossed nodes as well as over the uncrossed nodes. The latter merely specify a weight for $\mathbf{l}_S$. For example, the Weyl group of the Lie algebra of $L_S \cap Q$ sits inside $W_{\mathbf{p}}$ and so defines a relative Hasse diagram $W_{\mathbf{p}}^{\mathbf{q}}$ which we identify as a subset of $W_{\mathbf{p}}$ (and hence of $W_{\mathbf{g}}$) by representing each right coset by its unique minimal length representative. Relative Hasse diagrams may easily be calculated by adapting the method given above for ordinary Hasse diagrams.

In summary, our notation is

$W_{\mathbf{g}}$	Weyl group of $\mathbf{g}$—associated with topology and resolutions on G/B.
$W_{\mathbf{p}}$	Weyl group of a reductive Levi factor of $\mathbf{p}$ and a subgroup of $W_{\mathbf{g}}$.
$W^{\mathbf{p}} = W_{\mathbf{p}} \backslash W_{\mathbf{g}}$	Hasse subgraph of $\mathbf{p}$—associated with G/P.
$W_{\mathbf{p}}^{\mathbf{q}} = W_{\mathbf{q}} \backslash W_{\mathbf{p}}$	Hasse diagram associated to the fibration $G/Q \to G/P$. It is a subgraph of $W_{\mathbf{p}}$ and hence of $W_{\mathbf{g}}$.

THE BOTT–BOREL–WEIL THEOREM

It is now time to introduce the first key ingredient in computing the Penrose transform for homogeneous spaces. This is the *Bott–Borel–Weil theorem* [24], which computes the global sheaf cohomology of irreducible homogeneous vector bundles using the Weyl group and its affine action on weights. We shall use it in the Penrose transform in a *relative* form to compute direct images—see section 5.3 below. For completeness, we give a very simple proof of the theorem in section 5.1 below; the proof is in keeping with the flavour of the book so far and depends on first explicitly computing cohomology for the $SL(2,\mathbf{C})$ case and then reducing the general case to this one. For simplicity, the theorem will from now on be referred to simply as BBW

Theorem (5.0.1). [Bott–Borel–Weil] *Let G be a simply connected complex semisimple Lie group and $P \subseteq G$ a parabolic subgroup. Suppose $\lambda \in \mathbf{h}^*$ is an integral weight for G dominant with respect to P. Consider the homogeneous bundle $\mathcal{O}_\mathbf{p}(\lambda)$ on G/P. Then:*

- *If λ is singular for $\mathbf{g}$, then*

$$H^r(G/P, \mathcal{O}_\mathbf{p}(\lambda)) = 0 \; \forall r.$$

- *If λ is non-singular for $\mathbf{g}$, then as a representation of G,*

$$H^{\ell(w)}(G/P, \mathcal{O}_\mathbf{p}(\lambda)) = E(w.\lambda)$$

for $w \in W_\mathbf{g}$ being the unique element such that $w.\lambda$ is dominant. All other cohomology vanishes. (Recall that $w.\lambda$ refers to the affine action of the Weyl group as described in section 4.2.)

Remark (5.0.2). Notice that if λ is dominant for **p** and $w.\lambda$ is dominant for **g** as in this theorem, then $w.\lambda + \rho$ is dominant for **g** and

$$w^{-1}(w.\lambda + \rho) = w^{-1}(w(\lambda + \rho)) = \lambda + \rho$$

is dominant for **p** whence w^{-1} must be in the Hasse subgraph $W^\mathbf{p}$. This is one reason for our concern with $W^\mathbf{p}$ in section 4.3. It is used implicitly in section 5.2 and explicitly in section 5.3.

5.1 A simple proof

We shall first give a proof for G/B where B is Borel. As is often the case in these matters, a general proof for G/P follows by considering the fibration $G/B \to G/P$.

The structure theory of semisimple Lie algebras follows from the example of $\mathbf{sl}(2, \mathbf{C})$: it is used as a building block in the investigation of a general algebra. A similar comment applies to generalized flag manifolds. In order to study G/B, choose $\alpha \in \mathcal{S}$ and let P_α be the parabolic subgroup of G corresponding to the subset $\{\alpha\} \subseteq \mathcal{S}$ as in section 2.2. Thus, there is fibration

$$\pi : G/B \to G/P_\alpha$$

whose fibre is the Riemann sphere $\mathbf{P_1}$ as a homogeneous space for $\mathrm{SL}(2, \mathbf{C})$:

$$\mathbf{P_1} = \times = \mathrm{SL}(2, \mathbf{C}) / \begin{pmatrix} * & * \\ 0 & * \end{pmatrix}.$$

This is a well-known trick (cf. Demazure's proof of Bott's theorem [37]). It is easy to prove the BBW theorem for $\mathrm{SL}(2, \mathbf{C})$. This reads:

$$H^0(\overset{k}{\times}) = \overset{k}{\bullet} \text{ for } k \geq 0$$

$$H^0(\overset{k}{\times}) = 0 \text{ for } k \leq -1$$

$$H^1(\overset{j}{\times}) = \overset{-j-2}{\bullet} \text{ for } j \leq -2$$

$$H^1(\overset{j}{\times}) = 0 \text{ for } j \geq -1.$$

In more familiar notation, $\overset{k}{\times} = \mathcal{O}(k)$ is the sheaf of functions homogeneous of degree k and $\overset{k}{\bullet} = \odot^k E$ for E being the self-representation of $\mathrm{SL}(2, \mathbf{C})$. Thus, the first two equations above are just a restatement of Liouville's theorem of one-variable complex analysis. The last two equations either may be verified by using the Mayer–Vietoris sequence and Laurent expansions for the cover

$$\mathbf{P_1} = (\mathbf{P_1} \setminus \{\text{south pole}\}) \cup (\mathbf{P_1} \setminus \{\text{north pole}\})$$

[73] or may be deduced from the first two equations as an extremely special case of Serre duality [140]

$$H^1(\overset{j}{\times})^* = H^0(\Omega^1 \otimes \overset{-j}{\times}) = H^0(\overset{-2}{\times} \otimes \overset{-j}{\times}) = H^0(\overset{-j-2}{\times}).$$

Notice that $\overset{j}{\bullet} \mapsto \overset{-j-2}{\bullet}$ is indeed the affine action of the Weyl group of $\mathrm{sl}(2,\mathbf{C})$, generated by just one simple reflection.

To use this in the general case consider the following typical example:

$$G/B = \times\!-\!\times\!\!\rangle\!\!-\!\!\times \overset{\pi}{\longrightarrow} \times\!-\!\bullet\!\!\rangle\!\!-\!\!\times = G/P_\alpha.$$

The $\mathbf{P_1}$ case above applies to the fibres of this mapping to allow computation of direct images:

$$\pi^0_*(\overset{a \quad k \quad c}{\times\!-\!\times\!\rangle\!-\!\times}) = \overset{a \quad k \quad c}{\times\!-\!\bullet\!\rangle\!-\!\times} \text{ if } k \geq 0 \tag{8}$$

$$\pi^0_*(\overset{a \quad k \quad c}{\times\!-\!\times\!\rangle\!-\!\times}) = 0 \text{ if } k \leq -1 \tag{9}$$

$$\pi^1_*(\overset{a \quad j \quad c}{\times\!-\!\times\!\rangle\!-\!\times}) = \overset{a+j+1 \quad -j-2 \quad c+2j+2}{\times\!-\!\bullet\!\rangle\!-\!\times} \text{ if } j \leq -2 \tag{10}$$

$$\pi^1_*(\overset{a \quad j \quad c}{\times\!-\!\times\!\rangle\!-\!\times}) = 0 \text{ if } j \geq -1. \tag{11}$$

The third of these equations comes from the affine action of the simple reflection σ_α inside the Weyl group $W_{\mathbf{g}}$. In particular, note that for any irreducible representation $\overset{a \quad b \quad c}{\times\!-\!\times\!\rangle\!-\!\times}$ of B, at most one direct image is non-zero. Thus, the Leray spectral sequence [69] degenerates to give, for $j \leq -2$,

$$H^p(\overset{a \quad j \quad c}{\times\!-\!\times\!\rangle\!-\!\times}) = H^{p-1}(G/P_\alpha, \pi^1_*(\overset{a \quad j \quad c}{\times\!-\!\times\!\rangle\!-\!\times}))$$

$$= H^{p-1}(G/P_\alpha, \overset{a+j+1 \quad -j-2 \quad c+2j+2}{\times\!-\!\bullet\!\rangle\!-\!\times})$$

$$= H^{p-1}(G/P_\alpha, \pi^0_*(\overset{a+j+1 \quad -j-2 \quad c+2j+2}{\times\!-\!\times\!\rangle\!-\!\times}))$$

$$= H^{p-1}(\overset{a+j+1 \quad -j-2 \quad c+2j+2}{\times\!-\!\times\!\rangle\!-\!\times}).$$

To summarize, for α corresponding to the middle node of the Dynkin diagram $\bullet\!-\!\bullet\!\rangle\!-\!\bullet$, if λ is an integral weight with $\langle \lambda, \alpha^\vee \rangle \leq -2$, then

$$H^p(G/B, \mathcal{O}_\mathbf{b}(\lambda)) = H^{p-1}(G/B, \mathcal{O}_\mathbf{b}(\sigma_\alpha.\lambda)).$$

Notice also that if λ is singular by virtue of this α^{th} node, that is if $\langle \lambda, \alpha^\vee \rangle = -1$, then both direct images vanish under $\pi : G/B \to G/P_\alpha$ and so all cohomology $H^p(G/B, \mathcal{O}_\mathbf{b}(\lambda))$ vanishes.

Of course, these are perfectly general results, i.e., they hold for any complete flag manifold and any simple root.

Lemma (5.1.1). *Suppose that λ is an integral weight for the simply connected complex semisimple Lie group G and consider the cohomology of the holomorphic line bundle $\mathcal{O}_{\mathbf{b}}(\lambda)$ on G/B. If λ is singular, then all cohomology vanishes. If λ is non-singular and λ lies to the non-dominant side of the αth root plane for some simple positive root α (in other words, $\langle \lambda, \alpha^{\vee} \rangle \leq -2$), then*

$$H^p(G/B, \mathcal{O}_{\mathbf{b}}(\lambda)) = H^{p-1}(G/B, \mathcal{O}_{\mathbf{b}}(\sigma_\alpha.\lambda)).$$

Repeated application of this lemma now shows that if $w \in W_{\mathbf{g}}$ is chosen so that $w.\lambda$ is dominant for $\mathbf{g}$, then

$$H^{p+\ell(w)}(G/B, \mathcal{O}_{\mathbf{b}}(\lambda)) = H^p(G/B, \mathcal{O}_{\mathbf{b}}(w.\lambda))$$

and

$$H^p(G/B, \mathcal{O}_{\mathbf{b}}(\lambda)) = 0 \text{ for } p < \ell(w).$$

However, the lemma also shows that if λ is dominant, then all higher cohomology vanishes. Specifically, if $w_0 \in W_{\mathbf{g}}$ denotes the longest element, then

$$H^p(G/B, \mathcal{O}_{\mathbf{b}}(\lambda)) = H^{p+\ell(w_0)}(G/B, \mathcal{O}_{\mathbf{b}}(w_0.\lambda))$$

but $\ell(w_0) = \dim_{\mathbf{C}}(G/B)$ and so the right-hand side vanishes for $p \geq 1$. To summarize:

Theorem (5.1.2). [Bott] *Suppose that λ is an integral weight for the simply connected complex semisimple Lie group G and consider the cohomology of the holomorphic line bundle $\mathcal{O}_{\mathbf{b}}(\lambda)$ on G/B. If λ is singular, then all cohomology vanishes. If λ is non-singular and $w \in W_{\mathbf{g}}$ is the unique element such that $w.\lambda$ is dominant, then*

$$H^{\ell(w)}(G/B, \mathcal{O}_{\mathbf{b}}(\lambda)) = \Gamma(G/B, \mathcal{O}_{\mathbf{b}}(w.\lambda))$$

and cohomology in all other dimensions vanishes.

The Borel–Weil theorem [24] now identifies

$$\Gamma(G/B, \mathcal{O}_{\mathbf{b}}(\lambda)) = E(\lambda)$$

for dominant λ. However, this too may be proved by similar means as follows. In chapter 8 we shall construct a resolution of the constant sheaf $E(\lambda)$

on G/B by sheaves of holomorphic sections of homogeneous line bundles. Actually, we shall only sketch the proof of the existence of this Bernstein–Gelfand–Gelfand resolution but the first step is straightforward and gives an exact sequence

$$0 \to E(\lambda) \to \mathcal{O}_{\mathbf{b}}(\lambda) \to \bigoplus_{\alpha \in \mathcal{S}} \mathcal{O}_{\mathbf{b}}(\sigma_\alpha.\lambda).$$

Now by Bott's theorem above, the right-hand term has no global sections so

$$E(\lambda) \stackrel{\cong}{\to} \Gamma(G/B, \mathcal{O}_{\mathbf{b}}(\lambda)).$$

Finally, we say how to deduce the BBW for G/P. Consider the projection

$$\tau : G/B \to G/P.$$

The fibre of τ is itself homogeneous: as described in section 2.4, the fibres may be identified with P/B which is isomorphic to $L_S/(L_S \cap B)$ for L_S the semisimple part of L, the reductive part of P. Hence, since $L_S \cap B$ is a Borel subgroup of L_S, these fibres are complete generalized flag manifolds and the BBW theorem now applies fibrewise. In particular, if λ is an integral weight for G, non-singular and dominant for P, then

$$\mathcal{O}_{\mathbf{p}}(\lambda) = \tau_*^0(\mathcal{O}_{\mathbf{b}}(\lambda))$$

and all higher direct images vanish. The Leray spectral sequence degenerates to an isomorphism

$$H^r(G/P, \mathcal{O}_{\mathbf{p}}(\lambda)) = H^r(G/B, \mathcal{O}_{\mathbf{b}}(\lambda))$$

and the full BBW theorem is immediate.

5.2   Some examples

To compute examples, we use $W^{\mathbf{p}}$ as given in the examples of section 4.3 above, for various $\mathbf{p}$.

Example (5.2.1).   [Borel–Weil]
If λ is already **g**-dominant,

$$H^0(G/P, \mathcal{O}_{\mathbf{p}}(\lambda)) \cong E(\lambda)$$

and all other cohomology is zero.

Example (5.2.2).   [Hyperplane section bundle]

If $G/P = \times\!\!-\!\!\bullet\!\!-\!\!\bullet \cdots \bullet\!\!-\!\!\bullet = \mathbf{CP_n}$ then for

$$\lambda = \overset{k}{\bullet}\!\!-\!\!\overset{0}{\bullet}\!\!-\!\!\overset{0}{\bullet} \cdots \overset{0}{\bullet}\!\!-\!\!\overset{0}{\bullet}$$

we have that $\mathcal{O}_\mathbf{p}(\lambda) \cong \mathcal{O}(k)$. For $k \geq 0$, it follows from the previous example that

$$H^0(\mathbf{CP_n}, \mathcal{O}(k)) \cong \overset{k}{\bullet}\!\!-\!\!\overset{0}{\bullet}\!\!-\!\!\overset{0}{\bullet} \cdots \overset{0}{\bullet}\!\!-\!\!\overset{0}{\bullet} = \odot^k((\mathbf{C}^n)^*).$$

This isomorphism is easily realized in *abstract index notation* [127,128]. If $Z^\alpha \in \mathbf{C}^{n+1}$ so that $[Z^\alpha]$ are homogeneous coordinates for $\mathbf{CP_n}$ and if $A_{\alpha\ldots\beta} = A_{(\alpha\ldots\beta)} \in \odot^k((\mathbf{C}^n)^*)$, then the isomorphism above is given by

$$A_{(\alpha\ldots\beta)} \mapsto A_{\alpha\ldots\beta} Z^\alpha \ldots Z^\beta.$$

If $-n \leq k < 0$, we can readily check that $\lambda + \rho$ is singular. For if the simple roots are ordered from left to right in the Dynkin diagram for A_n, then $\sigma_{\alpha_{-k+1}} \cdots \sigma_{\alpha_2} \sigma_{\alpha_1} (\lambda + \rho)$ has a zero over the $(-k)^{\text{th}}$ node. Thus, in this range, all cohomology vanishes.

If $k \leq -n - 1$, then $H^n(G/P, \mathcal{O}(k)) \cong \overset{0}{\bullet}\!\!-\!\!\overset{0}{\bullet}\!\!-\!\!\overset{0}{\bullet} \cdots \overset{0}{\bullet}\!\!-\!\!\overset{-k-n-1}{\bullet}$

$$\cong \odot^{-k-n-1}(\mathbf{C}^{n+1}).$$

These results are standard in complex analysis [63].

Example (5.2.3).   [Global vector fields]

Now consider Minkowski space $\bullet\!\!-\!\!\times\!\!-\!\!\bullet$. Since $\overset{1}{\bullet}\!\!-\!\!\overset{0}{\times}\!\!-\!\!\overset{1}{\bullet}$ is dominant already, we have

$$H^i(\mathbf{M}, \overset{1}{\bullet}\!\!-\!\!\overset{0}{\times}\!\!-\!\!\overset{1}{\bullet}) = \begin{cases} 0 & \text{if } i \neq 0 \\ \overset{1}{\bullet}\!\!-\!\!\overset{0}{\bullet}\!\!-\!\!\overset{1}{\bullet} = \mathbf{sl}(4, \mathbf{C}) & \text{if } i = 0. \end{cases}$$

Now $\overset{1}{\bullet}\!\!-\!\!\overset{0}{\times}\!\!-\!\!\overset{1}{\bullet}$ is the holomorphic tangent bundle of Minkowski space (see example 3.2.3) and since $\mathrm{SL}(4, \mathbf{C})$ acts on Minkowski space by left translation there is an evident map

$$\mathbf{sl}(4, \mathbf{C}) \to \Gamma(\mathbf{M}, \overset{1}{\bullet}\!\!-\!\!\overset{0}{\times}\!\!-\!\!\overset{1}{\bullet})$$

whose image consists of the conformal Killing vector fields on $\mathbf{M}$. The BBW theorem tells us that this is an isomorphism and the only global holomorphic vector fields on $\mathbf{M}$ are the Killing fields.

Indeed, if θ is a highest root of $\mathbf{g}$ for any $\mathbf{g}$ then

$$H^i(G/P, \mathcal{O}_{\mathbf{p}}(\theta)) = \begin{cases} 0 & \text{if } i \neq 0 \\ \mathbf{g} & \text{if } i = 0. \end{cases}$$

Recall that the holomorphic tangent bundle Θ of G/P is induced by $\mathbf{g}/\mathbf{p}$; the $\mathbf{p}$-module with lowest weight $-\theta$ is a quotient of this and it follows that $\mathcal{O}_{\mathbf{p}}(\theta)$ is a quotient of Θ. If $\mathbf{p} = \mathbf{l} \oplus \mathbf{u}$ is a Levi decomposition of $\mathbf{p}$, then the remaining composition factors in Θ are of the form $\mathcal{O}_{\mathbf{p}}(\mu)$ where μ is a root of $\mathbf{u}$ with $\mu \prec \theta$ so that at least one nodal coefficient of μ is strictly negative. It follows that $\mu + \rho$ is not dominant regular so that $H^0(G/P, \mathcal{O}_{\mathbf{p}}(\mu)) = 0$. Taking the long exact sequence in cohomology for each short exact sequence in a composition series for Θ now shows that

$$H^i(G/P, \Theta) = \begin{cases} 0 & \text{if } i \neq 0 \\ \mathbf{g} & \text{if } i = 0. \end{cases}$$

So again the natural map

$$\mathbf{g} \to \Gamma(G/P, \Theta)$$

is an isomorphism—otherwise put, the global vector fields on G/P are precisely the infinitesimal symmetries of G/P.

Example (5.2.4).   [Cohomology on Minkowski space]
Consider •——×——•, again. As a first example of a bundle with no non-trivial global cohomology at all consider $\lambda = $ $\overset{0}{\bullet}\!\!-\!\!\overset{-2}{\times}\!\!-\!\!\overset{0}{\bullet}$. Then $\sigma_1(\lambda + \rho) = \sigma_1\left(\overset{1}{\bullet}\!\!-\!\!\overset{-1}{\times}\!\!-\!\!\overset{1}{\bullet}\right) = \overset{-1}{\bullet}\!\!-\!\!\overset{0}{\times}\!\!-\!\!\overset{1}{\bullet}$ so that λ is singular and so

$$H^i(\mathbf{M}, \overset{0}{\bullet}\!\!-\!\!\overset{-2}{\times}\!\!-\!\!\overset{0}{\bullet}) = 0 \quad \text{for every } i.$$

Sections of this line bundle are metrics in the conformal class on Minkowski space so we see in particular that none of these metrics is globally defined.

Next let $\lambda = \overset{1}{\bullet}\!\!-\!\!\overset{-2}{\times}\!\!-\!\!\overset{1}{\bullet}$ so that $\mathcal{O}_{\mathbf{p}}(\lambda) = \Omega^1$. Then

$$\sigma_2(\lambda + \rho) = \sigma_2\left(\overset{2}{\bullet}\!\!-\!\!\overset{-1}{\times}\!\!-\!\!\overset{2}{\bullet}\right) = \overset{1}{\bullet}\!\!-\!\!\overset{1}{\times}\!\!-\!\!\overset{1}{\bullet} = \rho$$

so that

$$H^1(\mathbf{M}, \Omega^1) = \mathbf{C}$$

and all other cohomology vanishes. The Kähler form given by the projective embedding of the following chapter generates this isomorphism.

More generally, the reader may verify that each irreducible summand of Ω^p on $\mathbf{M}$ has p^{th} cohomology equal to $\mathbf{C}$ whilst the cohomology in all other degrees vanishes. For instance, $\Omega^3 = \overset{1 \quad -4 \quad 1}{\bullet\!\!-\!\!\times\!\!-\!\!\bullet}$ and we compute

$$\sigma_1\sigma_3\sigma_2 \cdot \overset{1 \quad -4 \quad 1}{\bullet\!\!-\!\!\times\!\!-\!\!\bullet} = \overset{0 \quad 0 \quad 0}{\bullet\!\!-\!\!\times\!\!-\!\!\bullet}$$

so that

$$H^3(\mathbf{M}, \Omega^3) = \mathbf{C}.$$

More examples of how to use the BBW theorem to compute cohomology appear in the next section.

5.3   Direct images

The main use of the Bott–Borel–Weil theorem in this book is to compute higher direct images of holomorphic homogeneous sheaves under fibrations between generalized flag manifolds. Consider, for $Q \subset P$ standard parabolic,

$$G/Q \overset{\tau}{\to} G/P.$$

Then, as observed in section 2.4, the fibre P/Q is itself a generalized flag manifold. If $\mathbf{l}_S$ is the semisimple part of the reductive Levi factor $\mathbf{l}$ of $\mathbf{p}$ and if L_S is the corresponding subgroup of G, then $P/Q \cong L/(L \cap Q) \cong L_S/(L_S \cap Q)$. Applying the BBW theorem along these fibres to compute the direct image of $\mathcal{O}_{\mathbf{q}}(\lambda)$ on G/P, we seek a $\mathbf{p}$-dominant weight in the orbit

$$(W_{\mathbf{p}}^{\mathbf{q}})^{-1}.\lambda \equiv \{w^{-1}.\lambda \text{ s.t. } w \in W_{\mathbf{p}}^{\mathbf{q}}\}.$$

Recall that, as in section 4.4, $W_{\mathbf{p}}^{\mathbf{q}}$ is the Hasse subgraph of $W_{\mathbf{p}}$ associated to the parabolic $\mathbf{q} \cap \mathbf{l}_S$ and regarded as a subgraph of $W_{\mathbf{g}}$ by means of the inclusion $W_{\mathbf{l}_S} = W_{\mathbf{p}} \subseteq W_{\mathbf{g}}$. If none of these weights is $\mathbf{p}$-dominant (i.e., λ is $\mathbf{p}$-singular), then *all* direct images of $\mathcal{O}_{\mathbf{q}}(\lambda)$ vanish. Otherwise there is a unique $\mathbf{p}$-dominant element, $w^{-1}.\lambda$, and the direct image $(R^{\ell(w)}\tau_*)\mathcal{O}_{\mathbf{q}}(\lambda)$, or simply $\tau_*^{\ell(w)}\mathcal{O}_{\mathbf{q}}(\lambda)$, is given by $\mathcal{O}_{\mathbf{p}}(w^{-1}.\lambda)$ whereas all the other direct images vanish.

Recall that the method of determining $W_{\mathbf{p}}^{\mathbf{q}}$ as in sections 4.3 and 4.4 directly gives $(W_{\mathbf{p}}^{\mathbf{q}})^{-1}$ instead. This makes it especially straightforward to compute direct images in the Dynkin diagram notation:

Recipe for computing direct images

Step one: Determine the Hasse diagram $W_{\mathbf{p}}^{\mathbf{q}}$ by allowing $W_{\mathbf{p}}$ to act on $\rho^{\mathbf{q}}$. It is only necessary to record the *simple* reflections.

Step two: To compute the direct images of $\mathcal{O}_{\mathbf{q}}(\lambda)$ add one to each node coefficient of λ (so forming $\lambda + \rho$) and act on the result (as one acted on $\rho^{\mathbf{q}}$) with the graph of simple reflections constructed in step one.

Step three: If any element of the resulting orbit is repeated, λ is **p**-singular and all direct images vanish. Otherwise ...

Step four: Precisely one element of the orbit has positive entries over all nodes not crossed through in **p**. Subtract one from each of its node coefficients obtaining μ, say. This is **p**-dominant. If ℓ is the number of simple reflections required to produce μ, then

$$\tau_*^{\ell} \mathcal{O}_{\mathbf{q}}(\lambda) = \mathcal{O}_{\mathbf{p}}(\mu)$$

and all other direct images vanish.

Remark (5.3.1).   In practice, of course, it is seldom necessary to compute the entire orbit in steps one and two. Indeed, step one can be omitted entirely! This is because, if, in step two, we start with a weight $\lambda + \rho$ which has large negative numbers over all the uncrossed nodes and act on this with the simple reflections of $W_{\mathbf{p}}$, then we will create precisely the same pattern as acting on $\rho^{\mathbf{q}}$. In other words, we will construct $(W_{\mathbf{p}}^{\mathbf{q}})^{-1}$. For a weight which is not so antidominant, the pattern persists as far as the first occurrence of a weight dominant for **p**. At this stage, however, if the weight is not strictly dominant (i.e., if there is a zero over one of the uncrossed nodes), then all direct images vanish (in other words, step three is valid) whereas, if the weight is strictly dominant, the fourth step can be invoked. This should become clear in the following:

Example (5.3.2).   Consider ×—•—•⇒× → ×—•—•⇒•, numbering the simple root associated to the ith node α_i. So we are studying the projection of the projectivized bundle of pure spinors on $\mathbf{CS}^9$ to $\mathbf{CS}^9$. Apply the recipe to compute direct images of (i) $\overset{0\ \ 0\ \ 0\ \ -3}{×—•—•⇒×}$ and (ii) $\overset{0\ \ 0\ \ 0\ \ -8}{×—•—•⇒×}$.

Step one: Compute $W_{\mathbf{p}}^{\mathbf{q}}$ by acting on $\overset{0\ \ 0\ \ 0\ \ 1}{×—•—•⇒×}$ with simple reflections corresponding to the nodes α_2, α_3 and α_4. Obtain the following graph, for use in the next step:

$$\overset{(4)}{\underset{×—•—•⇒×}{0\ 0\ 0\ 1}} \to \overset{(3)}{\underset{×—•—•⇒×}{0\ 0\ 1\ \text{-}1}} \to \overset{}{\underset{×—•—•⇒×}{0\ 1\ \text{-}1\ 1}} \overset{\nearrow \searrow}{} \cdots$$

(4) $\overset{0\ 1\ 0\text{-}1}{×—•—•⇒×}$ (2)

(2) $\overset{1\ \text{-}1\ 0\ 1}{×—•—•⇒×}$ (4)

$\cdots \overset{\nearrow \searrow}{} \overset{}{\underset{×—•—•⇒×}{1\ \text{-}1\ 1\ \text{-}1}} \to \overset{(3)}{\underset{×—•—•⇒×}{1\ 0\ \text{-}1\ 1}} \to \overset{(4)}{\underset{×—•—•⇒×}{1\ 0\ 0\ \text{-}1}}$

where (i) denotes σ_{α_i}.

Step two: (i) Consider $\lambda = \overset{0\ \ \ 0\ \ \ 0\ \ \ -3}{\times\!\!-\!\!\bullet\!\!-\!\!\bullet\!\!\Rightarrow\!\!\times}$. Add ones to the node coefficients to obtain $\overset{1\ \ \ 1\ \ \ 1\ \ \ -2}{\times\!\!-\!\!\bullet\!\!-\!\!\bullet\!\!\Rightarrow\!\!\times}$. Act with the simple reflections in the graph of the previous step, obtaining

$$
\overset{(4)}{\underset{1\,1\,1\text{-}2}{\times\bullet\bullet\times}} \to
\overset{(3)}{\underset{1\,1\text{-}1\,2}{\times\bullet\bullet\times}} \to
\overset{(4)\,1\,0\,1\,0\,(2)}{\underset{1\,0\,1\,0}{\times\bullet\bullet\times}}
\begin{matrix}\nearrow\\\searrow\end{matrix}
\quad
\begin{matrix}\searrow\\\nearrow\end{matrix}
\overset{(3)}{\underset{1\,0\,1\,0}{\times\bullet\bullet\times}} \to
\overset{(4)}{\underset{1\,1\text{-}1\,2}{\times\bullet\bullet\times}} \to
\overset{(4)}{\underset{1\,1\,1\text{-}2}{\times\bullet\bullet\times}}.
$$

$$
\overset{(2)\,1\,0\,1\,0\,(4)}{}
$$

Note the zeros which appear after just two reflections and which ensure repetitions.

(ii) Consider $\lambda = \overset{0\ \ \ 0\ \ \ 0\ \ \ -8}{\times\!\!-\!\!\bullet\!\!-\!\!\bullet\!\!\Rightarrow\!\!\times}$. Add ones to the coefficients over the nodes to obtain $\overset{1\ \ \ 1\ \ \ 1\ \ \ -7}{\times\!\!-\!\!\bullet\!\!-\!\!\bullet\!\!\Rightarrow\!\!\times}$. Act with the simple reflections in the graph of the previous step, obtaining

$$
\overset{(4)}{\underset{1\,1\,1\text{-}7}{\times\bullet\bullet\times}} \to
\overset{(3)}{\underset{1\,1\text{-}6\,7}{\times\bullet\bullet\times}} \to
\overset{(4)\,1\text{-}5\,1\,5\,(2)}{\underset{1\text{-}5\,6\text{-}5}{\times\bullet\bullet\times}}
\begin{matrix}\nearrow\\\searrow\end{matrix}
\quad
\begin{matrix}\searrow\\\nearrow\end{matrix}
\overset{(3)}{\underset{\text{-}4\,5\text{-}4\,5}{\times\bullet\bullet\times}} \to
\overset{(4)}{\underset{\text{-}4\,1\,4\text{-}3}{\times\bullet\bullet\times}} \to
\overset{}{\underset{\text{-}4\,1\,1\,3}{\times\bullet\bullet\times}}.
$$

$$
\overset{(2)\,\text{-}4\,5\,1\text{-}5\,(4)}{}
$$

Step three: In (i), all direct images vanish.

Step four: In (ii), the *last* element in the orbit yields

$$
\tau^6_* \mathcal{O}_q\!\left(\overset{0\ \ \ 0\ \ \ 0\ \ \ -8}{\times\!\!-\!\!\bullet\!\!-\!\!\bullet\!\!\Rightarrow\!\!\times}\right) = \overset{-5\ \ \ 0\ \ \ 0\ \ \ 2}{\times\!\!-\!\!\bullet\!\!-\!\!\bullet\!\!\Rightarrow\!\!\bullet}.
$$

6

REALIZATIONS OF G/P

To conclude our survey of some of the geometry of generalized flag manifolds, we shall present two realizations of these manifolds using standard techniques from algebraic and symplectic geometry. We shall do so at this point because we now have the requisite machinery in place. But we earnestly recommend that the reader leave this chapter for a second reading, since it is only incidental to the rest of the book. Hasten on to the next chapter, where we begin the Penrose transform!

The first of these, the *n-tuple embedding*, realizes a G/P as a *complex projective variety*; indeed, it is easy to give a direct embedding into the projective space of a finite dimensional G-module. This shows, amongst other things, that G/P is a Kähler manifold, as claimed. The imaginary part of the Kähler structure is a symplectic structure, preserved by the action of the compact real form G_0 of G. Consequently, we may take the *moment map* of this action and so obtain a second realization of G/P as an orbit in the co-Adjoint representation of G_0.

Closely related to these realizations is the study of orbit structures on G/P. There are two possible kinds of orbit that might be studied. The first are those which arise from the left action on G/P of a Borel subgroup of G. These are easy to describe—they are all affine and affinely embedded and in one-to-one correspondence with the elements of W^P. The dimension of an orbit corresponds to the length of an element and the partial order on the graph W^P indicates when an orbit is in the closure of one of higher dimension. Indeed, all of this can be realized explicitly and we do this below. The set of orbits *stratifies* G/P; the stratification is *perfect* and it determines the integral homology of the space. It is also the link between the structure of G/P and the structure of a large class of **g**-modules. The link is realized in the beautiful geometry of Kempf [98] used in the proof by Beilinson–Bernstein [15] and Brylinski–Kashiwara [28] of the *Kazhdan–Lusztig conjecture* [97].

The example which may be familiar to the mathematical physicist is complexified compactified Minkowski space, viewed as a generalized flag manifold as above, and acted on by the lower triangular matrices of SL(4,**C**). This can be constructed as the union of an affine Minkowski space (**C**^{4}—the largest orbit) and the light cone "at infinity". This light cone itself is a union of five orbits—the vertex ι_0, a single generator (not including the vertex), two totally null two-surfaces (less their intersection in the distinguished generator), and the remainder of the cone. Each orbit is biholomorphic with some **C**^{k}. This example is quite generic—note especially how the boundaries of orbits can be singular.

The second possibility is to study orbits of a non-compact real form G_u of G on G/P. These are, unfortunately, rather more difficult to describe and we shall not do so here—the interested reader may consult the exhaustive work of Wolf [160] on the subject. Again this geometry relates to some representation structure theory, namely the form given by Vogan [151] for the *Langlands* classification of *Harish Chandra* modules and the Kazhdan–Lusztig conjecture in that setting. It is also extremely important in the construction of unitary representations of G_u; the simplest form of this is the technique of geometric quantization [77,102,163]. An advanced version of geometric quantization, using L^2-cohomology and finally constructed in [136,137], obtains the so-called *discrete series* of unitary representations for G_u. The problem is to construct an invariant inner product on a Harish Chandra module. There are indications that this can be done *without* L^2-cohomology by using the Penrose transform (in a guise sometimes called the *twistor transform*). (More on this, again, in chapters 10 and 11.)

The example of Minkowski space will again be familiar to the mathematical physicist. Here, G_u is SU(2,2). Thinking of Minkowski space as the Grassmannian Gr$_2$(**C**^{4}), again, the orbits of G_u consist of sets of planes on which the restriction of the Hermitian form has a particular signature and (in)definiteness. The possibilities are $(+,+)$, $(-,-)$, $(+,-)$, $(0,+)$, $(0,-)$, and $(0,0)$. The last of these is the (unique) closed orbit of SU(2,2) which is just *real* compactified Minkowski space. The orbits $(+,+)$ and $(-,-)$ are usually referred to as **M**$^+$ and **M**$^-$; see, for example, [127].

The material we present is standard—further details may be found in [18,77,160]. Much of what we say is not immediately related to the construction of the Penrose transform; it seems reasonable to include it since it is both extremely beautiful and readily accessible given the material we have presented thus far. We shall indicate in detail how G/P may be realized as a projective variety and how the Weyl group $W_{\mathbf{g}}$ can be used to compute the cohomology ring $H^*(G/P,\mathbf{C})$. Also, we shall explicitly construct a moment mapping $G/P \to \mathbf{g}_0^*$ to realize G/P as a co-Adjoint orbit.

6.1   The projective realization

In section 4.3 we assigned a weight $\rho^{\mathbf{p}}$ to a parabolic subalgebra $\mathbf{p}$ of $\mathbf{g}$ by letting

$$\langle \rho^{\mathbf{p}}, \alpha^{\vee} \rangle = \begin{cases} 0 & \alpha \in \mathcal{S}_{\mathbf{p}} \\ 1 & \alpha \in \mathcal{S} \setminus \mathcal{S}_{\mathbf{p}}. \end{cases}$$

Take G to be simply connected, and consider the finite dimensional irreducible G-representation $F(\rho^{\mathbf{p}})$ with highest weight $\rho^{\mathbf{p}}$. Let f be a highest weight vector in $F(\rho^{\mathbf{p}})$ (unique up to scale) and consider the action of G on f. Being maximal, f is annihilated by the positive roots spaces of $\mathbf{g}$ and hence fixed by their exponential group. Maximality and $\langle \rho^{\mathbf{p}}, \alpha^{\vee} \rangle = 0$ for $\alpha \in \mathcal{S}_{\mathbf{p}}$, hence for $\alpha \in \Delta(\mathbf{l})$, implies that f is fixed by the reductive Levi factor L of P, also. The Cartan subgroup H (corresponding to $\mathbf{h}$) preserves f up to scale, and so it follows that P is contained in the stabilizer of the line $[f]$ in the projective representation of G on $\mathbf{P}(F(\rho^{\mathbf{p}}))$. It is easy to see that if $\mathbf{g} = \mathbf{u} \oplus \mathbf{p}$ and if $y \in \mathbf{u}_-$, say $y \in \mathbf{g}_{\alpha}$ for $\alpha \in \Delta(\mathbf{u}_-)$, then $\langle \rho^{\mathbf{p}}, \alpha^{\vee} \rangle < 0$, so that y acts non-trivially on $[f]$. y cannot kill $[f]$ because, for $x \in \mathbf{g}_{-\alpha}$,

$$xyf = [x, y]f = \langle x, y \rangle \langle \rho^{\mathbf{p}}, \alpha \rangle f \neq 0.$$

Thus, P is precisely the stabilizer of the line $[f]$, and

$$G/P \cong G[f] \subset \mathbf{P}(F(\rho^{\mathbf{p}})),$$

realizing G/P projectively.

Remark (6.1.1).   By the Borel–Weil theorem, $E(\rho^{\mathbf{p}}) = \Gamma(G/P, \mathcal{O}_{\mathbf{p}}(\rho^{\mathbf{p}}))$. The above construction is then nothing more than the *n-tuple embedding* construction of algebraic geometry. A similar construction obviously works for any line bundle $\mathcal{O}_{\mathbf{p}}(\lambda)$ provided $\lambda - \rho^{\mathbf{p}}$ has no negative nodes. Note that the pull-back of the tautological bundle on $\mathbf{P}(F(\rho^{\mathbf{p}}))$ under such a realization is just $\mathcal{O}_{\mathbf{p}}(\lambda)$.

Example (6.1.2).   [Complex spheres] Consider Minkowski space $\bullet\!\!-\!\!\times\!\!-\!\!\bullet$; then $\rho^{\mathbf{p}} = \overset{0}{\bullet}\!\!-\!\!\overset{1}{\bullet}\!\!-\!\!\overset{0}{\bullet}$. $F(\rho^{\mathbf{p}})$ is just $\wedge^2 \mathbf{C}^4$ with $\mathrm{SL}(4, \mathbf{C})$ acting on $\mathbf{C}^4$ by the self-representation. Alternatively, $F(\rho^{\mathbf{p}})$ is the self-representation of $\mathrm{SO}(6, \mathbf{C})$. It is easy to see that a highest weight vector for this representation is a *simple* bivector, equivalently *null*, as a vector in the self-representation of $\mathrm{SO}(6, \mathbf{C})$, and that its orbit under G simply consists of all such. So the above construction giving the projective embedding finds $\bullet\!\!-\!\!\times\!\!-\!\!\bullet \hookrightarrow \mathbf{CP}^5$ and identifies $\bullet\!\!-\!\!\times\!\!-\!\!\bullet$ as the quadric of null directions at the origin of $\mathbf{C}^6$, as we saw in example 2.3.2. Similarly, the projective construction above applied to the complexified spheres $\mathbf{CS}^p$, that is to and , realizes each as the

projective light cone of the origin in $\mathbf{C}^{p+2}$.

Example (6.1.3).   [Pure spinors] The associated representation for the space

$$\mathbf{Z}^{2n} = \;\; \text{}$$

is a (reduced) spinor representation for $\mathrm{Spin}(2n+2, \mathbf{C})$. A highest weight vector is easily seen to be a *pure* spinor (see, for example, [81]) and its orbit again consists of all such. So Z is the space of projective pure spinors as claimed in example 2.3.5 and is therefore a space of totally null n-planes in the quadric $\mathbf{CS}^{2n}$. It is the *natural* higher dimensional analogue of Penrose's original (projective) twistor space. Similar comments apply to the other higher dimensional twistor spaces.

6.2   The cell structure of G/P

We now indicate how an affine stratification of G/P may be constructed which characterizes the integral homology of the space. A little further work, using the full Weyl group $W_{\mathbf{g}}$, characterizes the cohomology ring also. A model for this is the decomposition of $\mathbf{CP}^n$ into a disjoint union of a point and one copy of $\mathbf{C}^k$ for $1 \leq k \leq n$. For example, $\mathbf{CP}^2$ is $\mathbf{C}^2$ with a $\mathbf{CP}^1$ "at infinity" which is itself stratified as the one point compactification of $\mathbf{C}$.

For $w \in W_{\mathbf{g}}$ let f^w be a vector of weight $w^{-1}\rho^{\mathbf{P}}$ in $F(\rho^{\mathbf{P}})$ defined up to scale, so the projective point $[f^w]$ is well defined. As in chapter 4, the extremal weights of $F(\rho^{\mathbf{P}})$ are in one-to-one correspondence with the elements of $W^{\mathbf{P}}$. Thus we obtain, for each $w \in W^{\mathbf{P}}$, a distinct point $[f^w]$ in the projective realization of G/P. Pick a weight vector basis $\{f^i\}$ of $F(\rho^{\mathbf{P}})$, and let $\{f_i\}$ be the dual basis of $F(-\rho^{\mathbf{P}})$. Let $U = \exp \mathbf{u}$ in G where $\mathbf{p} = \mathbf{l} \oplus \mathbf{u}$ is the Levi decomposition as above. Then, for $w \in W^{\mathbf{P}}$, define the *Schubert cell*

$$X_w = \{\, [f] \in U[f^w] \text{ s.t } f_w(f) \neq 0 \,\}.$$

X_w is an affine subvariety of G/P open in its closure, the *Schubert variety* $\overline{X_w}$. To see this, note that the annihilator of $[f^w]$ in U corresponds to the root spaces in $\mathbf{u}$ conjugate to positive root spaces under w. For under such a conjugation of $\mathbf{g}$, $[f^w]$ becomes a highest weight space. It follows that X_w is coördinatized by $w\mathbf{u} \cap w^0\mathbf{u}$ (where w^0 is the longest element of $W^\mathbf{P}$ so that $w^0\mathbf{u}$ is a direct sum of negative root spaces). Hence X_w has dimension $\ell(w)$ and is contained in the open affine "big cell" $ww^0 U[f^w]$. The boundary

$$\overline{X_w} \setminus X_w = \overline{X_w} \cap \{\, [f]; f_w(f) = 0 \,\}$$

(which is generally a *singular* variety) contains exactly those $[f^{w'}]$ with $w' \prec w$. This is evident if we think of "raising" $[f^{w^0}]$ through the extremal weight spaces of $F(\rho^\mathbf{P})$ using $\mathbf{u}$, repeatedly. The same consideration shows that $G/P = \cup_{w\in W^\mathbf{P}} X_w$. In summary, we have proved:

Lemma (6.2.1). G/P is stratified by affinely embedded affine cells X_w, of dimension $\ell(w)$ for $w \in W^\mathbf{P}$ and

$$\partial \overline{X_w} = \bigcup_{w'\in W^\mathbf{P};\, w' \prec w} X_{w'}.$$

Corollary (6.2.2).   The integral homology module of G/P is freely generated by the Schubert varieties $\overline{X_w}$.

In other words, we have the remarkable fact that the directed graph $W^\mathbf{P}$ records the homology of G/P.

Remark (6.2.3).   The *Bruhat decomposition* of G with respect to P [156] is the disjoint union $G = \amalg_{w\in W^\mathbf{P}} BwP$, for the (presently fixed) Borel subgroup B; so the strata X_w are just the B orbits on G/P.

Remark (6.2.4).   For each $w \in W^\mathbf{P}$ and for each $U_- = \exp \mathbf{u}_-$, where $\mathbf{g} = \mathbf{u}_- \oplus \mathbf{p}$, $wU_-[f^w]$ is an affine coordinate system about $[f^w]$, again referred to as a "big cell".

Example (6.2.5).   [Even dimensional quadrics] For Minkowski space, $W^\mathbf{P}$ was computed in example 4.3.2 above. The reader will easily relate this to the cell decomposition given in the introduction to this chapter. For even-dimensional quadrics, the situation is similar ($W^\mathbf{P}$ is given in example 4.3.7). Thus, their integral homology is zero in odd degrees, $\mathbf{Z} \oplus \mathbf{Z}$ in the middle degree, and $\mathbf{Z}$ in all other degrees.

Example (6.2.6).   [Projective spaces and odd dimensional quadrics] The Hasse diagram associated to projective space $\mathbf{CP}^n$ has a single element of length ℓ for $0 \le \ell \le n$ (see example 4.3.6); the resulting stratification is the usual one by copies of $\mathbf{C}^\ell$ "at infinity" so that the integral homology of $\mathbf{CP}^n$ is $\mathbf{Z}$ in even degrees [25].

On the other hand, the Hasse diagram for an odd dimensional quadric similarly has a single element in each length (see example 4.3.8); so an odd dimensional sphere is a homology projective space even though these spaces are quite distinct. For example, the line bundle $\overset{1\ \ 0\ \ 0\ \ \ \ 0\ \ 0}{\times\!\!-\!\!-\!\!\bullet\!\!-\!\!\bullet\ \cdots\ \bullet\!\!\Rightarrow\!\!\bullet}$ has a $(2n+2)$-dimensional space of sections whereas no such a line bundle exists on $\mathbf{CP}^n$. Thus they are *holomorphically* distinct. We shall see in a moment that they are *topologically* distinct by computing the ring structure on their cohomology. Then the difference is unsurprising, for the ring structure depends on the full Weyl group $W_{\mathbf{g}}$.

So $W^{\mathbf{P}}$ does *not*, in general, completely determine the topology of G/P; notice, however, that it is occasionally possible for the same manifold to admit distinct homogeneous structures. This certainly occurs whenever G can be embedded in a larger group G' so that U maps to a unipotent subgroup U' which is the unipotent factor of a parabolic $P' \subset G'$, for then $G'/P' \cong G/P$.

Example (6.2.7).   Take $G = \mathrm{Sp}(2n)$ and let $G' = \mathrm{SL}(2n)$. Then G acts transitively on $\mathbf{CP}^{2n-1}$ identifying it with the space $\times\!\!-\!\!\bullet\!\!-\!\!\bullet\ \cdots\ \bullet\!\!\Leftarrow\!\!\bullet$. This space is the homogeneous complex *contact* manifold associated to $\mathrm{Sp}(2n)$ [21], obtained by projectivizing $\mathbf{C}^{2n}$ with a fixed *symplectic* structure preserved by $\mathrm{Sp}(2n)$.

Example (6.2.8).   Consider the exceptional quotients $\times\!\!\Rrightarrow\!\!\bullet$ and $\bullet\!\!\Rrightarrow\!\!\times$; both are five dimensional and for both the Hasse diagrams have a single element of each length ℓ, $0 \le \ell \le 5$, so that both spaces have the same homology groups as $\mathbf{CP}^5$. They are not holomorphically $\mathbf{CP}^5$ since the line bundles $\overset{1\ \ 0}{\times\!\!\Rrightarrow\!\!\bullet}$ and $\overset{0\ \ 1}{\bullet\!\!\Rrightarrow\!\!\times}$ have fourteen and seven dimensional spaces of global sections. Actually it is easy to identify the second space directly. The representation $F(\overset{0\ \ \ 1}{\bullet\!\!\Rrightarrow\!\!\bullet})$ used in its projective realization is the restriction of the self-representation of $\mathrm{SO}(7)$ under the inclusion $G_2 \hookrightarrow \mathrm{SO}(7)$; it follows that $\bullet\!\!\Rrightarrow\!\!\times$ is the five quadric. The space $\times\!\!\Rrightarrow\!\!\bullet$ is topologically different, again, as we shall see in the next section.

6.3 Integral cohomology rings

To say more about the topology of G/P, we shall investigate its integral cohomology. As a module, of course, it is simply the dual of the integral homology. But its cup product ring structure is more delicate. We shall investigate this by first computing the first Chern classes of homogeneous line bundles on G/P and then computing their products, and so the total Chern class of any homogeneous bundle. Central to this computation is the observation that since we may regard $W^{\mathbf{p}}$ as a subgraph of $W_{\mathbf{g}}$, each Schubert variety of G/P lifts to a Schubert variety of G/B. It follows that if $\nu : G/B \to G/P$ is the projection, then $\nu^* : H^*(G/P, \mathbf{Z}) \to H^*(G/B, \mathbf{Z})$ is an injection which maps a basis dual to the Schubert varieties in G/P to a similar one on G/B[1].

So let us compute the first Chern class of a homogeneous line bundle $\mathcal{O}_{\mathbf{p}}(\lambda_i)$. Recall the exact sequence of sheaves

$$0 \to \mathbf{Z} \overset{\times 2\pi i}{\to} \mathcal{O} \overset{\exp}{\to} \mathcal{O}^\times \to 0,$$

where $\mathcal{O}^\times$ is the sheaf of nowhere-vanishing holomorphic functions on G/P. $H^1(G/P, \mathcal{O}^\times)$ is the space of line bundles on G/P and (using the BBW theorem) the long exact sequence on cohomology gives an isomorphism

$$H^1(G/P, \mathcal{O}^\times) \overset{\mathrm{ch}_1}{\to} H^2(G/P, \mathbf{Z}) \cong \mathbf{Z}^{k(\mathbf{p})} \qquad (12)$$

where $k(\mathbf{p}) = |\mathcal{S} \setminus \mathcal{S}_{\mathbf{p}}|$. The image of a line bundle under ch_1 is its *first Chern class*, and the isomorphism with $\mathbf{Z}^{k(\mathbf{p})}$ is obtained by pairing with Schubert varieties $\overline{X_w}$ with $\ell(w) = 1$ and $w \in W^{\mathbf{p}}$. To calculate the composition, use:

Lemma (6.3.1).   [18] *Let $\alpha \in \mathcal{S} \setminus \mathcal{S}_{\mathbf{p}}$ and let σ_α be its simple reflection. Then*

$$\mathrm{ch}_1(\mathcal{O}_{\mathbf{p}}(\lambda_i))[\overline{X_{\sigma_\alpha}}] = \langle \lambda_i, \alpha \rangle.$$

In particular, in (12) *$\mathbf{Z}^{k(\mathbf{p})}$ is the lattice in $H^2(G/P, \mathbf{C}) \subset H^2(G/B, \mathbf{C}) = \mathbf{h}^*$ spanned by weights λ_i annihilating $[\mathbf{p}, \mathbf{p}] \cap \mathbf{h}$.*

(See remark 6.4.2 for an explicit calculation which can be used to prove this lemma.)

Let $x_i(\mathbf{p}) = \mathrm{ch}_1 \mathcal{O}_{\mathbf{p}}(\lambda_i)$—these Chern classes span $H^2(G/P, \mathbf{Z})$. By the remark above, there is no confusion in identifying $x_i(\mathbf{p})$ with $x_i(\mathbf{b})$ and writing x_i for either. Note that the $x_i(\mathbf{b})$ generate $H^*(G/B, \mathbf{Q})$ as a ring over $\mathbf{Q}$ but do *not*, in general, generate the integral cohomology ring.

The next step is actually to compute the ring structure on $H^*(G/B, \mathbf{Z})$ from which, again by the remark above, the ring structure of $H^*(G/P, \mathbf{Z})$ follows. Let $x_{(i}x_j \ldots x_{k)}$ (ℓ terms) be the symmetrized product in $H^\ell(G/B, \mathbf{Z})$. Then

---

[1]Indeed [18], $H^*(G/P, \mathbf{Z})$ is the space of $W_{\mathbf{p}}$-invariants in $H^*(G/B, \mathbf{Z})$.

Lemma (6.3.2).   [18] *If $w \in W_{\mathbf{g}}$,*

$$x_{(i}x_j \ldots x_k)[\overline{X_w}] = \sum \langle \lambda_{(i}\mu_1^\vee \rangle \langle \lambda_j \mu_2^\vee \rangle \cdots \langle \lambda_{k)}\mu_\ell^\vee \rangle,$$

where the sum runs over all collections $\mu_1, \ldots, \mu_\ell$ of positive roots of $\mathbf{g}$ such that

$$w^{-1} = \sigma_{\mu_1}\sigma_{\mu_2}\ldots\sigma_{\mu_\ell}$$

is a reduced expression. (Recall that these collections may be read off the graph structure of $W_{\mathbf{g}}$.)

Example (6.3.3).   We compute the initial part of the ring structure on $H^*(\text{⋈⟹}, \mathbf{Z})$ as follows: let $\lambda_1 = \overset{1 \quad 0}{\text{⟺}}$ and $\lambda_2 = \overset{0 \quad 1}{\text{⟺}}$. The initial part of the Weyl group of $\overset{\alpha_1 \quad \alpha_2}{\text{⟺}}$ is

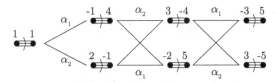

(the entry against each edge signifies the simple root whose corresponding reflection links the nodes—the remaining edges correspond to reflections in non-simple walls). If μ is not a simple root, then there exists a $w \in W_{\mathbf{g}}$ with $\alpha = w\mu$ simple—w is easily read from the Weyl group. Because reflection preserves $\langle \ , \ \rangle$, we have $\langle \lambda_i, \mu^\vee \rangle = \langle w^{-1}\lambda_i, \alpha^\vee \rangle$ and so these quantities are easily calculated. We find that

$$x_1^2[\overline{X_{(21)}}] = 3 \qquad x_1^2[\overline{X_{(12)}}] = 0$$
$$x_2^2[\overline{X_{(21)}}] = 0 \qquad x_2^2[\overline{X_{(12)}}] = 1$$
$$x_{(1}x_{2)}[\overline{X_{(21)}}] = 1 \quad x_{(1}x_{2)}[\overline{X_{(12)}}] = 1$$

where X_{ij} is the Schubert cell for $\sigma_{\alpha_i}\sigma_{\alpha_j}$. Observe that

$$x_1^2 + 3x_2^2 - 3x_1 x_2 = 0$$

(see [2]). From this, the initial algebra of $H^*(\text{⋈⟹}, \mathbf{Z})$ and $H^*(\text{⟹⋈}, \mathbf{Z})$ is easily derived. Note that on ⋈⟹, x_1^2 does not generate the fourth cohomology ($3 \neq 1$). This compares with the situation for $\mathbf{CP}^5$ where the first Chern class of the hyperplane section bundle generates the integral cohomology ring. So ⋈⟹ is not homeomorphic to $\mathbf{CP}^5$.

Finally we calculate the total Chern class of a homogeneous bundle $\mathcal{E}$ on G/P. To do this, we employ the *splitting principle* [25]. The bundle $\nu^*\mathcal{E}$

on G/B has a composition series whose terms are line bundles—evidently, there is a term $\mathcal{O}_b(\mu)$ for each weight μ (with multiplicity) in the **p**-module inducing $\mathcal{E}$. Then $\nu^*\mathrm{ch}(\mathcal{E})$ is the product of the $\mathrm{ch}(\mathcal{O}_b(\mu))$. Thus,

$$\mathrm{ch}_1(\mathcal{E}) = \text{formal character of representation inducing } \mathcal{E}.$$

Example (6.3.4).　　To illustrate, we compute the total Chern class and the first Pontrjagin class of the tangent bundles on ⟜ and ⟜. The weights of these two homogeneous bundles are easily read off the list of roots of G_2 and so, continuing the notation of the previous example,

$$\mathrm{ch}(\Theta(\text{⟜})) = (1 + 2x_1 + 3x_2)(1 + x_1 - x_2)(1 - x_2)(1 - x_1 + 3x_2)(1 + x_1)$$
$$\mathrm{ch}(\Theta(\text{⟜})) = (1 - x_1 + 2x_2)(1 + x_1 - x_2)(1 - x_2)(1 - x_1 + 3x_2)(1 + x_1).$$

Recall that if

$$\mathrm{ch}(\mathcal{E}) = \prod_{1 \le i \le n} (1 + d_i)$$

then

$$\mathrm{p}(\mathcal{E}) = \prod_{1 \le i \le n} (1 - d_i^2)$$

is the total Pontrjagin class of $\mathcal{E}$. The Pontrjagin class of a manifold is the Pontrjagin class of its tangent bundle and is a topological invariant. The first Pontrjagin class is its component in degree four. For example,

$$\mathrm{p}_1(\text{⟜}) = -1 \in \mathbf{Z} \cong H^4(\text{⟜}, \mathbf{Z})$$

compared with -6 for $\mathbf{CP}^5$, confirming, again, that these spaces are topologically distinct. Also,

$$\mathrm{p}_1(\text{⟜}) = -1 \in \mathbf{Z} \cong H^4(\text{⟜}, \mathbf{Z})$$

which is consistent with our identification of this space as $\mathbf{CS}^5$.

Very much more is known about the geometry of generalized flag manifolds explored using a projective realization and its relation to $W_{\mathbf{g}}$. See [2,18,22,85].

6.4　Co-Adjoint realizations and moment maps

The second realization of G/P is as an orbit in the dual $\mathbf{g}_0^*$ of the Lie algebra of the compact real form G_0 of G. This realization is important for the emphasis it places on the real structure of G/P, particularly its structure as a non-trivial example of a real symplectic manifold. We shall say more about the physical implications of this structure in a remark below. From a mathematical point of view, recall that any symplectic manifold with a symplectic action of G_0 is (a covering of) an orbit in $\mathbf{g}_0^*$; G/P is such a manifold by virtue of its projective realization (the symplectic structure following from the inherited Kähler structure and complex structure). The passage from the projective realization to the co-Adjoint realization is the *moment map* of the action of G_0 which we shall explicitly construct.

The first step is to observe that if $K_0 = G_0 \cap P$, then $G_0/K_0 \cong G/P$. Recall that it is possible to find a basis e_β of the root spaces of $\mathbf{g}$ so that the root space decomposition of $\mathbf{g}$ takes the form

$$\mathbf{g} = \mathbf{h} \oplus (\oplus_{\alpha \in \Delta^+} \mathbf{C}e_\alpha) \oplus (\oplus_{\alpha \in \Delta^+} \mathbf{C}e_{-\alpha}).$$

Then $\mathbf{g}_0$ may be taken to be

$$\mathbf{g}_0 = \mathbf{t} \oplus (\oplus_{\alpha \in \Delta^+} \mathbf{R}(e_\alpha - e_{-\alpha})) \oplus (\oplus_{\alpha \in \Delta^+} \mathbf{R}i(e_\alpha + e_{-\alpha}))$$

where $\mathbf{t}$ corresponds to $i\mathbf{h}_{\mathbf{R}}^*$ under the Killing form. $\mathbf{k}_0$ has a similar form where α ranges over the positive roots of a Levi factor of $\mathbf{p}$ and its complement in this direct sum will be denoted by $\wp_0$. $\mathbf{k}, \wp$ are the complexifications of these. It is clear from the differential action of $\wp_0$ that G_0 acts locally transitively so that the orbit of any point is open and closed (by compactness) and hence is all of G/P.

The second is to find a vector in $\mathbf{g}_0^*$ whose stabilizer under the co-Adjoint action of G_0 is K_0. The evident candidate is a weight canonically associated to $\mathbf{p}$; we have already encountered $\rho^{\mathbf{p}}$ defined by

$$\langle \rho^{\mathbf{p}}, \alpha^\vee \rangle = \begin{cases} 0 & \text{if } \alpha \in \mathcal{S}_{\mathbf{p}} \\ 1 & \text{if } \alpha \in \mathcal{S} \setminus \mathcal{S}_{\mathbf{p}}. \end{cases}$$

The infinitesimal complex co-Adjoint action of a vector $v \in \mathbf{g}_\alpha$ on $\rho^{\mathbf{p}}$ is given by

$$\text{coad}(v) \cdot \rho^{\mathbf{p}}(w) = -\langle \rho^{\mathbf{p}}, \alpha^\vee \rangle \langle v, w \rangle$$

for $w \in \mathbf{g}$. From this formula and the decomposition of $\mathbf{g}_0$, it follows that the stabilizer of $i\rho^{\mathbf{p}} \in \mathbf{t}^*$ under G_0 is K_0. So its orbit in $\mathbf{g}_0^*$ is G/P.

A complex projective space is a real symplectic manifold whose symplectic form is constructed from the Fubini–Study metric b and the complex structure J by the formula $\omega(v, w) = b(Jv, w)$. If $G/P \hookrightarrow \mathbf{P}(F(\rho^{\mathbf{p}}))$ is a projective realization, then G/P is a complex submanifold of $\mathbf{P}(F(\rho^{\mathbf{p}}))$

and it follows that the restriction of ω to G/P makes G/P a real symplectic manifold. It may be assumed that the Fubini–Study metric is invariant under G_0 and so the action of G_0 on G/P is symplectic and that there is a *moment map* $\Phi : G/P \to \mathbf{g}_0^*$ [77].

This symplectic form and moment map may be explicitly constructed as follows. A left invariant two-form on G_0 may be defined by the formula

$$\Omega_g(v, w) = i\frac{([v, w]f, f)}{(f, f)},$$

where v, w are right invariant vector fields on G_0, $[f] \in G/P \subset \mathbf{P}(F(\rho^\mathbf{p}))$ and $(\cdot, \cdot)$ is a G_0 invariant Hermitian form on $F(\rho^\mathbf{p})$. Ω is exact; define

$$\Phi_g(v) = i\frac{(vf, f)}{(f, f)}$$

and observe that, since Φ is also left invariant,

$$d\Phi(u, v) = \Phi([u, v]) = \Omega(u, v)$$

for u, v right invariant. Now Ω descends to a two-form on G/P. To see this, let u, v be root vectors in $\mathbf{g}$ with $u \in \mathbf{k}$, realized as right invariant vector fields. By the left invariance of Ω it is no loss of generality to suppose that $[f]$ is a highest weight space. Then either $[u, v] \in [k, k]$, in which case $[u, v]f = 0$, or $[u, v] \in p$, in which case $[u, v]f$ and f lie in different orthogonal weight spaces. In either case, $\Omega(u, v) = 0$. The resulting closed form, ω, is the G_0 invariant symplectic form on G/P.

Recall that if (X, ω) is a symplectic manifold with a symplectic G_0 action, then ω may be pulled back to a left invariant closed form on G_0; this represents a class in the Lie algebra cohomology group $H^2(\mathbf{g}_0, \mathbf{R})$, which is the second cohomology group of the complex of left invariant differential forms on G_0 with the de Rham differential. It is a classical result that $\mathbf{g}_0$ semisimple implies $H^2(\mathbf{g}_0, \mathbf{R}) = 0$. The pulledback form is therefore the differential of a left invariant one-form on G_0 which defines a mapping $X \to \mathbf{g}_0$, the *moment mapping* of the action [77,163]. In the present situation, this construction is transparent and the moment map is

$$\Phi : gP \to \Phi_g$$

(acting on $\mathbf{g}_0$ realized as the right invariant vector fields on G_0).

Observe that if $g \in G_0$, then

$$\Phi(g[f])v = i\frac{(v \cdot gf, gf)}{(f, f)} = i\frac{(g^{-1}vg \cdot f, f)}{(f, f)} = \text{Coad}(g)\Phi([f])v,$$

so that Φ intertwines the actions of G_0. If $[f]$ is a λ-weight space in $F(\rho^\mathbf{P})$, then the orthogonality of weight spaces implies that

$$\Phi([f])v = i\lambda(v).$$

In particular, if $[f]$ is the highest weight space of $F(\rho^\mathbf{P})$ then $\Phi([f]) = i\rho^\mathbf{P}$. In other words, the image of the moment map on the projective realization of G/P is its co-Adjoint realization.

Example (6.4.1).   To illustrate this construction, consider Minkowski space $\mathbf{M} \subset \mathbf{P}(T^{[\alpha\beta]})$ in the abstract index notation [127]. Let $Z^\alpha \rightsquigarrow \overline{Z}_\alpha$ [128] be the conjugate linear mapping $T^\alpha \to T_\alpha$ given by a Hermitian structure (of signature $(+, +, +, +)$) on T^α; so

$$(Z^\alpha, Z^\beta) = Z^\alpha \overline{Z}_\alpha.$$

Then $g_0 = so(6, \mathbf{R})$ and consists of $A^{[\alpha\beta]}_{[\mu\nu]}$ satisfying a reality condition:

$$\overline{A}^{[\mu\nu]}_{[\alpha\beta]} = \epsilon^{\mu\nu\sigma\tau}\epsilon_{\alpha\beta\gamma\delta}A^{[\gamma\delta]}_{[\sigma\tau]}$$

and a skew symmetry condition:

$$A^{[\alpha\beta]}_{[\mu\nu]} = -\epsilon_{\mu\nu\sigma\tau}\epsilon^{\alpha\beta\gamma\delta}A^{[\sigma\tau]}_{[\gamma\delta]}.$$

Here, $\epsilon^{\alpha\beta\gamma\delta}$ is the preserved element of $\wedge^4 T^\alpha$ under SL(4,$\mathbf{C}$); equivalently, regarding a skew pair $[\alpha\beta]$ as a single abstract index and $T^{[\alpha\beta]}$ as the self-representation of $so(6, \mathbf{C})$, ϵ is the preserved inner product. Its appearance in the two previous equations therefore serves to raise and lower such indices.

Then the moment map

$$\Phi : \mathbf{M} \to so(6)^*$$

is determined by

$$\Phi([X])(A) = iA^{[\alpha\beta]}_{[\mu\nu]}X^{[\mu\nu]}\overline{X}_{[\alpha\beta]}/X^{[\gamma\delta]}\overline{X}_{[\gamma\delta]}$$

or, taking account of the skew symmetry of A,

$$\Phi([X]) = \frac{1}{2}i(X^{[\mu\nu]}\overline{X}_{[\alpha\beta]} - X_{[\alpha\beta]}\overline{X}^{[\mu\nu]})/X^{[\gamma\delta]}\overline{X}_{[\gamma\delta]}$$

(where $X_{\alpha\beta} = \epsilon_{\alpha\beta\gamma\delta}X^{\gamma\delta}$, etc.).

Equivalently, take $\mathbf{g_0} = su(4)$, a typical element of which is a trace-free matrix A_β^α satisying

$$\overline{A}_\beta^\alpha = -A_\beta^\alpha.$$

Then

$$\Phi([X]) = \frac{2i}{X^{[\gamma\delta]}\overline{X}_{[\gamma\delta]}}(X^{\alpha\gamma}\overline{X}_{\beta\gamma} - \delta_\beta^\alpha X^{[\gamma\delta]}\overline{X}_{[\gamma\delta]}).$$

Remark (6.4.2).   Observe that Φ does *not* descend to a one-form on G/P for it does not annihilate vectors parallel to $G_0 \to G/P$. It follows that ω may represent a non-trivial class of $H^2(G/P, \mathbf{R})$. Which class it is is easy to check; for each $\alpha \in \mathcal{S} \setminus \mathcal{S}_{\mathbf{p}}$, $\mathcal{O}_\alpha = \overline{X_{\sigma_\alpha}}$ is the orbit of the identity coset under the corresponding copy of $SU(2) \subset G_0$. Then

$$\langle[\omega], [\mathcal{O}_\alpha]\rangle = \int_{\mathcal{O}_\alpha} \omega.$$

Setting

$$x_\alpha = \frac{1}{\sqrt{2}}(e_{-\alpha} - e_\alpha) \quad \text{and} \quad y_\alpha = \frac{i}{\sqrt{2}}(e_{-\alpha} + e_\alpha)$$

whose projections span the tangent space of $\mathcal{O}_\alpha$ at eP and using the left invariance of ω, we obtain

$$\langle[\omega], [\mathcal{O}_\alpha]\rangle = \omega_{eP}(x_\alpha, y_\alpha) = \langle\rho^{\mathbf{p}}, \alpha^\vee\rangle.$$

So $[\omega] = \rho^{\mathbf{p}} \in H^2(G/P, \mathbf{R}) \subset h_{\mathbf{R}}^*$ (see lemma 6.3.1).

In particular, $[\omega]$ is integral and is the first Chern class of the line bundle $\mathcal{O}_{\mathbf{p}}(\rho^{\mathbf{p}})$ on G/P, which, we have observed, is the restriction of the tautological line bundle on $\mathbf{P}(F(\rho^{\mathbf{p}}))$.

More generally, let λ be any dominant integral weight satisfying

$$\langle\lambda, \alpha^\vee\rangle \begin{cases} =0 & \text{if } \alpha \in \mathcal{S}_{\mathbf{p}} \\ \geq 1 & \text{if } \alpha \in \mathcal{S} \setminus \mathcal{S}_{\mathbf{p}}. \end{cases}$$

Then $\mathcal{O}_{\mathbf{p}}(\lambda)$ is ample and the associated n-tuple embedding realizes $G/P \hookrightarrow \mathbf{P}(F(\lambda))$. Given this embedding, there is a G_0-symplectic form ω_λ on G/P, constructed as above. Thus, ω_λ is determined by the left invariant form on G_0 defined at the identity by

$$\Omega_\lambda(x_\alpha, y_\beta) = \begin{cases} 0 & \text{if } \alpha \neq \beta \\ i\langle\lambda, \alpha^\vee\rangle & \text{if } \alpha = \beta \end{cases}$$

with $\Omega_\lambda(x_\alpha, x_\beta) = \Omega_\lambda(y_\alpha, y_\beta) = 0$. Then a moment map realizes G/P as the co-Adjoint orbit through $i\lambda$ with the standard symplectic form.

Given the symplectic realization of G/P, with one of these symplectic forms ω_λ the projective realization may be recovered using *geometric quantization*. ω_λ determines an integral cohomology class $\lambda \in H^2(G/P, \mathbf{R})$ and so a smooth homogeneous line bundle. The complex structure on the orbit is given by declaring the complex span of the projections of the right invariant vector fields $\{x_\alpha - iy_\alpha\}$, $\alpha \in \Delta(\mathbf{u})$, to be the holomorphic tangent bundle. This is a Lagrangian distribution on G/P and so yields a holomorphic structure on the line bundle, recovering $\mathcal{O}_{\mathbf{p}}(\lambda)$. Then $H^0(G/P, \mathcal{O}_{\mathbf{p}}(\lambda)) = E(\lambda)$ with its G_0-invariant Hermitian form is the *quantum Hilbert space* associated to the physical system described by $(G/P, \omega_\lambda)$. Of course, $E(\lambda) = F(\lambda)^*$.

Much information about this physical system can be obtained from $\mathbf{P}(F(\lambda))$, for the physical observables are the eigenvalues of operators on this space. The projectively embedded G/P is recovered by seeking certain special states, called *coherent states*, which are, in a suitable sense, the "best" approximation to the original classical states, i.e., to the points of G/P. Precisely, $x \in G/P \subset \mathbf{g}_0^*$ determines a vector $v_x \in \mathbf{g}_0$ via the Killing form and so an Hermitian operator $\hat{x} = -iv_x \cdot \in \mathsf{End}(F(\lambda))$. Finding the best approximation in $\mathbf{P}(F(\lambda))$ to x corresponds to maximizing the (normalized) matrix element

$$\varphi_x([f]) = \frac{(\hat{x}f, f)}{(f, f)}.$$

The stationary points of φ_x correspond to the eigenspaces of $\hat{x}$. If $x = eP$, so that $x = i\lambda \in \mathbf{g}_0^*$, then $\hat{x} = h_\lambda \in \mathbf{h}$ and the eigenspaces of $\hat{x}$ are the weight spaces of $F(\lambda)$. If f is a μ-weight vector, then $\varphi_x([f]) = \langle \mu, \lambda^\vee \rangle$; it is a standard result that $\langle \mu, \mu \rangle \leq \langle \lambda, \lambda \rangle$ so that φ_x is a maximum at the highest weight space which is therefore the coherent state associated to eP. But

$$\varphi_{\mathrm{coAd}(g) \cdot x} = L_g^* \varphi \tag{13}$$

and so the map

$$x \in G/P \subset \mathbf{g}_0^* \to \text{coherent state associated to } x$$

recovers the projective embedding $G/P \hookrightarrow \mathbf{P}(F(\lambda))$ as the G_0 orbit of the highest weight space.

Remark (6.4.3). $\varphi_{x|G/P}$ is a perfect Morse function; its set of critical points is the Weyl orbit of x. From (13) observe that if v is a right invariant vector field on G_0 and pr v denotes its projection onto $G_0/K_0 \cong G/P$, then

$$d\varphi_x(\mathrm{pr}\ v) = \frac{([v,\hat{x}]f,f)}{(f,f)}$$

$$= i\frac{([v_x,v]f,f)}{(f,f)}$$

$$= \Omega(v_x,v)$$

so that the projection of $-v_x$, regarded as a right invariant vector field, is the Hamiltonian vector field associated to φ_x and the resulting cell decomposition of G/P is exactly that outlined above.

Remark (6.4.4).   It is amusing to note that the correspondence implied by a double fibration has a neat interpretation in terms of coherent states. Let $y \in G/R$ be realized as a point in the co-Adjoint orbit of $i\rho^r \in \mathbf{g}_0^*$. As above, y determines an Hermitian operator $\hat{y}$ on $F(\lambda)$. The coherent states in $\mathbf{P}(F(\rho^p))$ corresponding to y are those maximizing

$$\varphi_y([f]) = \frac{(\hat{y}f,f)}{(f,f)}.$$

In particular, restrict attention to those $[f] \in G/P$. (This restriction is called the *Hartree–Fock* approximation in physics and chemistry.) Again, suppose $y = eQ$ so that $\hat{y} = h_{\rho^p}$ and the critical submanifolds of φ are the intersections of direct sums of extremal weight spaces and G/P. At an extremal μ-weight space, $\varphi_y = \langle \mu, \rho^{q\vee}\rangle$ which is maximized for μ dominant, i.e., $\mu = \rho^p$. Recalling (13), it follows that φ_y is maximized on $Q \cdot eP$, i.e., on the subvariety of G/P corresponding to y, and that this last statement is true for arbitrary y.

Coherent states may be more generally defined in the setting of symplectic manifolds with suitable symplectic group actions; given two such manifolds, suitably restricted, the observation of this remark will yield a double fibration and attendant Penrose transform between the manifolds of coherent states. The details of this symplectic Penrose transform are yet to be investigated.

THE PENROSE TRANSFORM IN PRINCIPLE

It is now time to set up the *Penrose transform*—we shall do this first in its most general setting in this chapter and then, in chapter 9, we shall specialize to the homogeneous setting in which, thanks to the representation theory of the earlier chapters, it becomes computable.

The general form of the Penrose transform is as follows. Suppose X,Y and Z are complex manifolds and suppose we are given a double fibration

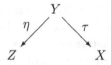

where η and τ are surjective mappings of maximal rank and the pair (η, τ) embeds Y as a submanifold of $Z \times X$. This shall be called a *correspondence* between Z and X. The simplest example is where the fibres of τ are just single points, whence Y is merely the graph of a function

$$\chi : X \longrightarrow Z$$
$$\uplus \qquad\qquad \uplus$$
$$x \longmapsto \eta(\tau^{-1}(x))$$

In this case the Penrose transform is simply the pull-back of functions

$$\chi^* : \Gamma(Z, \mathcal{O}) \to \Gamma(X, \mathcal{O}),$$

where $\chi^* f \equiv f \circ \chi$ for f a holomorphic function on Z. More generally, points of X describe a family of submanifolds of Z and vice versa:

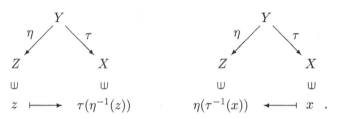

The situation is not symmetrical, however, because the fibres of τ are always supposed to be compact. The Penrose transform starts with data on Z in the form of an analytic cohomology class

$$\omega \in H^p(Z, \mathcal{O}(E))$$

for E a holomorphic bundle on Z. In its simplest form, the Penrose transform consists of just restricting this cohomology class to the submanifolds $\tau(\eta^{-1}(x))$ as $x \in X$ varies:

$$(\mathsf{P}\omega)(x) \equiv H^p(\tau(\eta^{-1}(x)), \mathcal{O}(E)).$$

Since, by assumption, Y may be regarded as a submanifold of $Z \times X$, each $\tau(\eta^{-1}(x))$ is isomorphic to the fibre $\eta^{-1}(x)$ and, in particular, is compact. Thus, $H^p(\tau(\eta^{-1}(x)), \mathcal{O}(E))$ is finite dimensional [34] and, if of constant dimension, gives a holomorphic vector bundle on X as $x \in X$ varies. In this way, $\mathsf{P}\omega$ has a natural interpretation as a section of this bundle. The sections which arise in this way are usually further restricted in some way, often as being annihilated by a holomorphic differential operator.

In order to investigate this rather naïve description of the transform, it is evidently a good idea to split it up into two steps.

- **Pull back to Y** Roughly speaking, the cohomology class ω is regarded as a cohomolgy class on Y which is constant up the fibres of η. This constancy is interpreted by means of a differential equation. If suitable topological conditions, to be described in section 7.1, pertain on these fibres, then this pull-back is an isomorphism.

- **Push down to X** Restricting ω to $\eta(\tau^{-1}(x))$ now coincides with the notion of direct image under τ. The transform may therefore be investigated by means of the Leray spectral sequence which is designed exactly for the situation. In this way the differential equations along the fibres of η manifest themselves on X and the transform may be proved to be an isomorphism onto an appropriate solution space.

In fact, this way of looking at things also deals with the general case of the Penrose transform where the simple idea of just restricting to the correspondence submanifolds is inappropriate. In other words, the Leray spectral sequence is able to interpret the data on X in all cases. In general, there is often a final step taken down on X where the output of the whole process is subject to some further reinterpretation. In the classical case (for Minkowski space), the simplest form is the twistor description of *right-handed fields* whereas the less obvious transform gives a description of *left-handed fields*. The combined form of the transform in the classical context is given in [44].

Our application of this machine is in the case in which X, Y and Z are corresponding open subvarieties in a *double fibration* of generalized flag varieties. If P and R are standard parabolic subgroups of G, then so is $Q = P \cap R$ and there is a double fibration

$$
\begin{array}{ccc}
 & G/Q & \\
{\scriptstyle \eta}\swarrow & & \searrow{\scriptstyle \tau} \\
G/R & & G/P
\end{array}
$$

(see section 2.4). We take X to be an open subset of G/P (most often an affine "big cell" or an open orbit of a real form G_u of G) and let $Y = \tau^{-1}X$ and $Z = \eta(Y)$. The transform is then applied to calculate the cohomology on Z of the restriction of irreducible homogeneous bundles on G/R.

7.1 Pulling-back cohomology

This section discusses just one aspect of the Penrose transform, namely the first step of pulling back cohomology classes on Z to Y by means of the mapping $\eta : Y \to Z$. For the purposes of this discussion, it is irrelevant as to whether Y and Z are homogeneous.

Given any holomorphic vector bundle E on Z with $\mathcal{O}(E)$ as the sheaf of holomorphic sections, there is a natural map on cohomology:

$$
\eta^{-1} : H^r(Z, \mathcal{O}(E)) \to H^r(Y, \eta^{-1}\mathcal{O}(E))
$$

where $\eta^{-1}\mathcal{O}(E)$ is the topological inverse image sheaf of $\mathcal{O}(E)$; in other words, the sections of $\eta^*(E)$ which are locally constant along the fibres of η. Certainly, if these fibres are connected, then there is an isomorphism on sections:

$$
H^0(Z, \mathcal{O}(E)) \xrightarrow{\cong} H^0(Y, \eta^{-1}\mathcal{O}(E))
$$

and this leads to the question as to what happens on higher cohomology.

To investigate this question, consider the Dolbeault resolution (e.g., [63,78,157])

$$0 \to \mathcal{O}(E) \to \mathcal{E}^{0,0}(E) \xrightarrow{\bar{\partial}} \mathcal{E}^{0,1}(E) \xrightarrow{\bar{\partial}} \mathcal{E}^{0,2}(E) \to \cdots$$

of $\mathcal{O}(E)$ where $\mathcal{E}^{p,q}(E)$ denotes the sheaf of smooth forms of type (p,q) with values in E. This gives rise to a resolution

$$0 \to \eta^{-1}\mathcal{O}(E) \to \eta^{-1}\mathcal{E}^{0,\bullet}(E)$$

of $\eta^{-1}\mathcal{O}(E)$ and the homomorphism above may be realized as the composition:

$$\begin{aligned} H^r(Z,\mathcal{O}(E)) &\cong H^r(\Gamma(Z,\mathcal{E}^{0,\bullet}(E))) \\ &\cong H^r(\Gamma(Y,\eta^{-1}\mathcal{E}^{0,\bullet}(E))) \quad \text{(if η has connected fibres)} \\ &\to H^r(Y,\eta^{-1}\mathcal{O}(E)). \end{aligned}$$

Thus, the question as to whether this composition is an isomorphism is reduced to the question as to whether $\eta^{-1}\mathcal{E}^{0,\bullet}(E)$ is an *acyclic* resolution of $\eta^{-1}\mathcal{O}(E)$. This question is no longer in the holomorphic category and may be rephrased as follows. Suppose $\eta : Y \to Z$ is a smooth mapping of maximal rank between smooth manifolds. Suppose F is a smooth complex vector bundle on Z. Let $\mathcal{E}(F)$ denote the sheaf of smooth sections of F. Under what circumstances is it the case that

$$H^r(Y,\eta^{-1}\mathcal{E}(F)) = 0 \ ?$$

As an example, consider the case when Z is a single point. Then F is trivial and may as well be $\mathbf{C}$. The sheaf $\eta^{-1}\mathcal{E}(F)$ is just the constant sheaf $\mathbf{C}$ and the vanishing of $H^r(Y,\eta^{-1}\mathcal{E}(F))$ is therefore precisely the vanishing of the de Rham cohomology $H^r(Y,\mathbf{C})$. The cohomology $H^r(Y,\eta^{-1}\mathcal{E}(F))$, in general, may be similarly realized as a fibrewise de Rham cohomology. Since F is locally trivial and since $H^s(Z,\mathcal{E}(F)) = 0$, by partition of unity, it is clear that the discussion is local on Z (which may as well be a disc) and that F is actually irrelevant. Thus, it suffices to demonstrate the vanishing of $H^r(Y,\eta^{-1}\mathcal{E})$.

Let $\mathcal{E}^1_\eta$ denote the sheaf of one-forms *relative* to the mapping η, i.e., dual to the *vertical* vectors (tangent to the fibres of η). Thus, there is a projection

$$\mathcal{E}^1 \to \mathcal{E}^1_\eta$$

and a *relative* exterior derivative d_η defined as the composition

$$\mathcal{E} \xrightarrow{d} \mathcal{E}^1 \to \mathcal{E}_\eta^1.$$

This extends to a resolution

$$0 \to \eta^{-1}\mathcal{E} \to \mathcal{E} \xrightarrow{d_\eta} \mathcal{E}_\eta^1 \xrightarrow{d_\eta} \mathcal{E}_\eta^2 \to \cdots$$

in the obvious manner. This is a resolution by acyclics and, hence,

$$H^r(Y, \eta^{-1}\mathcal{E}) \cong \frac{\ker d_\eta : \Gamma(Z, \mathcal{E}_\eta^r) \to \Gamma(Z, \mathcal{E}_\eta^{r+1})}{\operatorname{im} d_\eta : \Gamma(Z, \mathcal{E}_\eta^{r-1}) \to \Gamma(Z, \mathcal{E}_\eta^r)}.$$

If η were a fibration (in other words, if we could take $Y = Z \times F$ for some fibre F with η as projection onto the first factor), then this would be just the deRham cohomology depending on a parameter $z \in Z$. Thus: $H^r(Y, \eta^{-1}\mathcal{E})$ would vanish precisely when $H^r(F, \mathbf{C})$ did. This would always be the case if η had compact fibres. In the case of the Penrose transform, it is almost always the case that η has non-compact fibres but, nevertheless, is often a fibration. The following argument is also valid for the cases of interest (see chapter 9). By mimicking the usual proof of homotopy invariance of ordinary deRham cohomology, it follows that $H^r(Y, \eta^{-1}\mathcal{E})$ is invariant under homotopies of Y which preserve η. Thus,

Theorem (7.1.1).   *If the fibres of η are contractible by means of smooth homotopies varying smoothly with $z \in Z$, then $H^r(Y, \eta^{-1}\mathcal{E}) = 0$ $\forall r \geq 1$.*

The definitive answer is given by Buchdahl [31] who showed that:

Theorem (7.1.2).   *If, for each fibre Y_z,*

$$H^r(Y_z, \mathbf{C}) = 0 \text{ and } H^{r-1}(Y_z, \mathbf{C}) = 0,$$

this second condition being replaced by connectivity of Y_z if $r = 0$, then $H^r(Y, \eta^{-1}\mathcal{E}) = 0$.

He also gives an example to justify the condition $H^{r-1}(Y_z, \mathbf{C}) = 0$. In general Michael Singer [142,143] has proved a local relative universal coefficient theorem:

Theorem (7.1.3).   *There is an exact sequence of $\mathcal{E}$-modules on Z*

$$0 \to \mathcal{E}xt^1(\mathcal{H}_{p-1}(\eta), \mathcal{E}) \to \mathcal{H}^p(\eta) \to \mathcal{H}om(\mathcal{H}_{p-1}(\eta), \mathcal{E}) \to 0$$

where

$$\mathcal{H}^r(\eta) \equiv \eta_*^r(\eta^{-1}\mathcal{E})$$

is a local fibrewise cohomology, and

$$\mathcal{H}_r(\eta)_z = H_r(y_z, \mathbf{C}) \otimes_{\mathbf{C}} \mathcal{E}_z$$

is a local fibrewise homology.

He has used this theorem to investigate the Penrose transform when the fibres of η have non-trivial topology. This case is especially relevant to the twistor description of zero rest mass fields with *sources*, e.g., electromagnetic fields with charges [7].

In any case, as explained earlier, we now conclude

Theorem (7.1.4).   *If the fibres of $\eta : Y \to Z$ are contractible by means of a smooth homotopy which preserves η, then*

$$H^r(Z, \mathcal{O}(E)) \overset{\cong}{\to} H^r(Y, \eta^{-1}\mathcal{O}(E))$$

for all r.

7.2   Pushing-down cohomology

The aim of this section is to describe the interpretation of $H^r(Y, \eta^{-1}\mathcal{O}(E))$ down on X under the mapping $\tau : Y \to X$. To do this, consider the resolution

$$0 \to \eta^{-1}\mathcal{O}(E) \to \Omega_\eta^\bullet(E)$$

of $\eta^{-1}\mathcal{O}(E)$ by the locally free sheaves

$$\Omega_\eta^\bullet(E) \equiv \Omega_\eta^\bullet \otimes_\mathcal{O} \eta^*\mathcal{O}(E)$$

where Ω_η^p denotes the sheaf of relative holomorphic p-forms defined by analogy with the smooth version of section 7.1. Notice that $d_\eta : \Omega_\eta^p \to \Omega_\eta^{p+1}$ induces a well-defined operator

$$\nabla_\eta : \Omega_\eta^p(E) \to \Omega_\eta^{p+1}(E)$$

since $\eta^*(E)$ is naturally given by transition functions constant along the fibres of η. In other words, $\eta^* E$ is *flat* relative to η. The operator ∇_η is a *relative connection* in that it is determined by $\nabla_\eta : \Omega_\eta^0(E) \to \Omega_\eta^1(E)$ which satisfies a Leibnitz rule $\nabla_\eta(fs) = f\nabla_\eta(s) + d_\eta f \otimes s$. Conversely, if η has connected and simply connected fibres, then the vector bundles on Y which arise from pull back from Z are precisely those which admit a flat relative connection, i.e., one with vanishing relative curvature, so that the $\Omega_\eta^\bullet(E)$ is exact. See [50] and section 9.9 below for more details and a discussion as to how this structure yields the *Ward transform* under push-down to X. By general abstract nonsense [42,69], the cohomology $H^r(Y, \eta^{-1}\mathcal{O}(E))$

is determined by the cohomology $H^q(Y, \Omega_\eta^p(E))$. Specifically, there is a spectral sequence (a special case of the *hypercohomology* spectral sequence):

$$E_1^{p,q} = H^q(Y, \Omega_\eta^p(E)) \Longrightarrow H^{p+q}(Y, \eta^{-1}\mathcal{O}(E))$$

and, thus, it suffices to calculate $H^q(Y, \Omega_\eta^p(E))$, an easier task since the coefficient sheaves are locally free. Recall that τ is always supposed to have compact fibres. Thus, for any fixed p the direct images $\tau_*^t \Omega_\eta^p(E)$ are at least coherent [65,71,99] and often locally free. Certainly, if everything is homogeneous, then $\tau_*^t \Omega_\eta^p(E)$ is always locally free. Indeed, in this case direct images are always computable by Bott–Borel–Weil (as explained in section 5.3). In any case, the cohomology $H^q(Y, \Omega_\eta^p(E))$ is given by the *Leray* spectral sequence (e.g., [69]):

$$E_2^{s,t} = H^s(X, \tau_*^t \Omega_\eta^p(E)) \Longrightarrow H^{s+t}(Y, \Omega_\eta^p(E)).$$

We shall usually assume that X is Stein, in which case the higher cohomology of X vanishes by Cartan's theorem B [63,78,157]. Certainly, it is always possible to consider a polydisc neighbourhood X' of x in X, replacing Y by $Y' \equiv \tau^{-1}(X')$ and Z by $Z' \equiv \eta(Y')$. In this way, we can say that the Penrose transform is always valid locally.

7.3   A spectral sequence

The discussion so far is now fairly standard [50] but, as will be demonstrated in chapter 9 it is crucial for the homogeneous case to incorporate a variation by allowing an arbitrary resolution to replace the holomorphic relative de Rham sequence. We summarize the preceding discussion as follows:

Theorem (7.3.1).   *Suppose that the fibres of $\eta \colon Y \to Z$ are contractible by means of a smooth homotopy which preserves η. Let*

$$0 \to \eta^{-1}\mathcal{O}(E) \to \Delta_\eta^\bullet(E)$$

be a resolution of $\eta^{-1}\mathcal{O}(E)$ by locally free sheaves. Suppose X is Stein. Then there is a spectral sequence

$$E_1^{p,q} = \Gamma(X, \tau_*^q \Delta_\eta^p(E)) \Longrightarrow H^{p+q}(Z, \mathcal{O}(E)).$$

The Penrose transform uses this spectral sequence to interpret $H^r(Z, \mathcal{O}(E))$ in terms of differential equations on X. Just as for the relative de Rham sequence, the differentials $\delta_\eta \colon \Delta_\eta^p(E) \to \Delta_\eta^{p+1}(E)$ are, in practice, designed to be differential operators so that, in case $\tau_*^q \Delta_\eta^p(E)$ are locally free, we obtain differential operators

$$\tau_*^q \delta_\eta : \tau_*^q \Delta_\eta^p(E) \longrightarrow \tau_*^q \Delta_\eta^{p+1}(E)$$

which induce the differentials of the E_1-level of the spectral sequence. Similar comments apply to higher levels of the spectral sequence. It is in this way that the differential equations on X arise.

For the moment, however, things must be left at this somewhat vague stage whilst the proposed improved resolution $\Delta_\eta^\bullet(E)$ is constructed in the homogenous case outlined at the beginning of this chapter. That is, we shall seek an efficient replacement for the relative de Rham resolution of $\eta^{-1}\mathcal{O}(E)$ for the case in which $\eta : G/Q \to G/R$ and $\mathcal{O}(E) = \mathcal{O}_r(\lambda)$ is an irreducible homogeneous sheaf on G/R, restricted to $Z \subset G/R$. This replacement, the (dual) *Bernstein–Gelfand–Gelfand* resolution, is the subject of the following chapter, after which the Penrose transform will be brought to life with many examples and applications in chapter 9.

THE BERNSTEIN–GELFAND–GELFAND RESOLUTION

On a holomorphic n-manifold the de Rham sequence can be used to resolve the constant sheaf $\mathbf{C}$ by locally free $\mathcal{O}$-modules

$$0 \to \mathbf{C} \to \Omega^0 \xrightarrow{d} \Omega^1 \xrightarrow{d} \Omega^2 \to \cdots \to \Omega^n \to 0$$

where Ω^p denotes the sheaf of holomorphic p-forms. For a general manifold, this is the best we can hope for, but on a homogeneous manifold there is a resolution due to Bernstein, Gelfand, and Gelfand which is, in many ways, more efficient. We shall need a relative or fibrewise version of this to compute the Penrose transform. In order to construct it, we start with the deRham complex and try to modify it as follows.

8.1   A prototype

The first thing to observe is that if Y_1 and Y_2 are vector fields on a manifold X which annihilate a function f, then their Lie bracket $[Y_1, Y_2]$ also annihilates f. Thus, if a distribution $D \subset T$ (the tangent bundle of X) has the property that $[D, D] = T$ then in order that a function be constant it is sufficient that it be killed by sections of D. Dually, let $\Omega^1 \to \Delta^1$ be the corresponding projection on sheaves. The initial segment of the de Rham resolution may be replaced by the exact sequence

$$0 \to \mathbf{C} \to \Delta^0 \xrightarrow{\delta} \Delta^1$$

where $\Delta^0 \equiv \Omega^0$ and δ is the composition $\Delta^0 = \Omega^0 \xrightarrow{d} \Omega^1 \to \Delta^1$. On a general manifold, there is no natural choice for D but on a homogeneous manifold we can make a natural choice at the identity coset and then move this choice around under the action of the group.

Consider, as a prototype, the homogeneous space ×—×, i.e., the complete flag manifold of $SL(3, \mathbf{C})$. The tangent space T at the identity coset may be identified with the strictly lower triangular matrices

$$\begin{pmatrix} 0 & 0 & 0 \\ * & 0 & 0 \\ * & * & 0 \end{pmatrix}$$

and within this space is the natural subspace D of matrices of the form

$$\begin{pmatrix} 0 & 0 & 0 \\ * & 0 & 0 \\ 0 & * & 0 \end{pmatrix}.$$

The inclusion $D \subset T$ is a homomorphism of B-modules, for B the upper triangular matrices. Specifically,

$$\begin{array}{c} \overset{2 \quad -1}{\times\!\!-\!\!\times} \\ \oplus \\ \underset{-1 \quad 2}{\times\!\!-\!\!\times} \end{array} \subset T = \frac{\mathrm{sl}(3, \mathbf{C})}{\mathbf{b}}.$$

Thus, D gives rise to a homogeneous distribution on ×—× which is, however, not integrable since

$$\left[\begin{pmatrix} 0 & 0 & 0 \\ * & 0 & 0 \\ 0 & * & 0 \end{pmatrix}, \begin{pmatrix} 0 & 0 & 0 \\ * & 0 & 0 \\ 0 & * & 0 \end{pmatrix} \right] = \begin{pmatrix} 0 & 0 & 0 \\ 0 & 0 & 0 \\ * & 0 & 0 \end{pmatrix}.$$

This is exactly as desired and we can conclude that there is an exact sequence

$$0 \to \mathbf{C} \to \overset{0 \quad 0}{\underset{\times\!\!-\!\!\times}{}} \to \begin{array}{c} \overset{-2 \quad 1}{\times\!\!-\!\!\times} \\ \oplus \\ \underset{1 \quad -2}{\times\!\!-\!\!\times} \end{array}$$

which can be completed into a resolution of $\mathbf{C}$ as follows. The kernel of the projection $\Omega^1 \to \Delta^1$ is the line bundle $\overset{-1 \ -1}{\times\!\!-\!\!\times}$. Since ×—× is three dimensional, the Hodge *-isomorphism gives $\Omega^2 = \Omega^3 \otimes \left(\Omega^1\right)^*$. As a homogeneous bundle, $\Omega^3 = \overset{-2 \ -2}{\times\!\!-\!\!\times}$. Combining these facts gives an exact sequence

$$0 \to \begin{array}{c} \overset{-3 \quad 0}{\times\!\!-\!\!\times} \\ \oplus \\ \underset{0 \quad -3}{\times\!\!-\!\!\times} \end{array} \to \Omega^2 \to \overset{-1 \ -1}{\times\!\!-\!\!\times} \to 0.$$

Thus, there is the following commutative diagram.

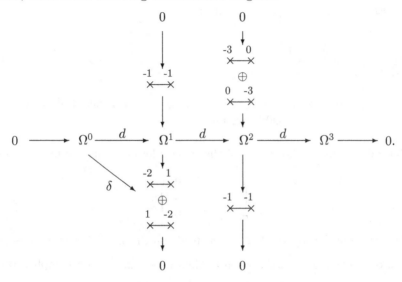

In this diagram, consider the composition

$$\overset{\text{-1 \ -1}}{\times\!\!-\!\!\times} \to \Omega^1 \overset{d}{\to} \Omega^2 \to \overset{\text{-1 \ -1}}{\times\!\!-\!\!\times} \ .$$

By the Leibnitz rule,

$$d(f\omega) = df \wedge \omega + f\, d\omega \quad \text{for } f \in \mathcal{O} \text{ and } \omega \in \Omega^1.$$

But, for $\omega \in \overset{\text{-1 \ -1}}{\times\!\!-\!\!\times}$,

$$df \wedge \omega \in \quad \begin{matrix} \overset{\text{-3 \ 0}}{\times\!\!-\!\!\times} \\ \oplus \\ \underset{\text{0 \ -3}}{\times\!\!-\!\!\times} \end{matrix} \quad \subset \Omega^2$$

so $\overset{\text{-1 \ -1}}{\times\!\!-\!\!\times} \to \overset{\text{-1 \ -1}}{\times\!\!-\!\!\times}$ is just linear over the functions, i.e., an invariant homomorphism of homogeneous bundles. Hence, by Schur's lemma, it is simply scalar multiplication. As demonstrated below, this scalar must be non-zero. Thus, the compositions

$$\Omega^1 \overset{d}{\to} \Omega^2 \to \overset{\text{-1 \ -1}}{\times\!\!-\!\!\times} \quad \text{and} \quad \overset{\text{-1 \ -1}}{\times\!\!-\!\!\times} \to \Omega^1 \overset{d}{\to} \Omega^2$$

split the vertical exact sequences in the above diagram. Notice that these vertical sequences certainly do *not* split as homogeneous bundles (or, indeed, as holomorphic bundles). Rather, they are split as **C**-sheaves by means of invariant differential operators. A simple diagram chase (i.e., pure thought) now establishes the exactness of the following *Bernstein–Gelfand–Gelfand* resolution:

It remains to be seen why $\overset{-1\ -1}{\times\!\!-\!\!\times} \to \overset{-1\ -1}{\times\!\!-\!\!\times}$ as constructed above is non-zero. There are two possible arguments.

Method 1   Label the entries of the strictly lower triangular matrices by

$$\begin{pmatrix} 0 & 0 & 0 \\ Y_1 & 0 & 0 \\ Y_3 & Y_2 & 0 \end{pmatrix}.$$

Thus, Y_j may be regarded as vector fields on $SL(3, \mathbf{C})$. Let ω_i denote the dual ome-forms: thus, $\omega_i(Y_j) = \delta_{ij}$. Then $\Omega^2 \to \overset{-1\ -1}{\times\!\!-\!\!\times}$ is accomplished on $SL(3, \mathbf{C})$ by $\theta \mapsto \theta(Y_1, Y_2)$. Also, $\omega_3 \in \overset{-1\ -1}{\times\!\!-\!\!\times} \subset \Omega^1$. Hence

$$(d\omega_3)(Y_1, Y_2) = \tfrac{1}{2}Y_1(\omega_3(Y_2)) - \tfrac{1}{2}Y_2(\omega_3(Y_1)) - \omega_3([Y_1, Y_2])$$

$$= 0 - 0 + \omega_3(Y_3)$$

$$= 1.$$

This calculation on $SL(3, \mathbf{C})$ shows that $\overset{-1\ -1}{\times\!\!-\!\!\times} \to \overset{-1\ -1}{\times\!\!-\!\!\times}$ down on $\times\!\!-\!\!\times$ is the identity.

Method 2   The inclusion $\overset{-1\ -1}{\times\!\!-\!\!\times} \hookrightarrow \Omega^1$ may be regarded as a one-form ω with coefficients in $\overset{1\ 1}{\times\!\!-\!\!\times}$. The kernel of ω is the distribution D and the statement that D is nowhere integrable is precisely equivalent to ω being a contact structure on $\times\!\!-\!\!\times$. But $\omega \wedge d\omega$ is precisely the homomorphism $\overset{-1\ -1}{\times\!\!-\!\!\times} \to \overset{-1\ -1}{\times\!\!-\!\!\times}$.

Remark (8.1.1).   Method 1 can be interpreted as a calculation down on $\times\!\!-\!\!\times$ using the affine coordinate patch of the cell structure described in chapter 6. This method evidently generalizes to allow calculation of similar operators on arbitrary complex flag manifolds. On the other hand, method 2 yields a resolution on any three-dimensional contact manifold and, indeed, may easily be modified [27] to yield an analogous resolution on contact manifolds of higher dimension.

Remark (8.1.2).   Even in the homogeneous case, however, it should be noticed that the claimed increase in efficiency has been achieved at some expense. The improvement over the de Rham sequence is in the dimensions of the bundles which in this case read

1 2 2 1   rather than   1 3 3 1.

The price to pay for this gain is that the differential operators are no longer necessarily first order. An increased order has crept in because the splittings used in the construction were also achieved by means of differential operators. By checking symbols, it follows that the four operators

are second order whilst the others are all first order.

Remark (8.1.3).   The usual discussion of the Bernstein–Gelfand–Gelfand resolution [17] employs a dual formulation in terms of Verma modules (see section 11.1 and theorem 11.2.1). The equivalence between these two viewpoints is discussed in [53]. As a sequence of Verma modules, the above resolution reads

and it is often useful to adopt this alternative formulation. From now on, either of these resolutions shall be known simply as BGG resolutions.

8.2   Translating BGG resolutions

An extremely useful device which may be applied to the BGG resolution is the *Jantzen–Zuckerman translation functor* (see, for example, [151]). The application of this functor on ✕—✕ is as follows. Choose any irreducible representation $\underset{\bullet\,\,\,\,\,\,\bullet}{\overset{p\,\,\,\,\,\,q}{\rule{0pt}{0pt}}}$ of $SL(3, \mathbf{C})$. As a representation of B (by restriction) this gives rise to a (reducible) homogeneous vector bundle on ✕—✕ which, as merely a vector bundle, is nothing other than the canonically trivial bundle

$\underset{\bullet\,\,\,\,\,\,\bullet}{\overset{p\,\,\,\,\,\,q}{\rule{0pt}{0pt}}} \times$ ✕—✕.

Thus, we may tensor the BGG resolution of $\mathbf{C}$ with this bundle to obtain a resolution of $\overset{p\quad q}{\bullet\!\!-\!\!\bullet}$. The essence of the Jantzen–Zuckerman functor is the fact that this resolution automatically decomposes according to central character (see [53] for some examples and chapter 11 for more Lie algebraic machinery which is relevant here). The result is the following BGG resolution:

At this point, the reader will recognize this diagram as a picture of the Weyl group of $SL(3, \mathbf{C})$ (see example 4.1.5) as a partially ordered set and, moreover, observe that the weights which appear are precisely those obtained by the affine action of the said Weyl group elements on the weight $\overset{p\quad q}{\bullet\!\!-\!\!\bullet}$. That the central characters agree is a consequence of Harish Chandra's theorem [94]. Indeed, conversely, this theorem implies (as in chapter 11 for example) that a homomorphism between Verma modules is necessarily restricted to those occurring in the same affine Weyl group orbit as above.

8.3   The general case on G/B

Bearing these observations in mind, it is easy to guess the general form of the BGG resolution on a complete flag manifold G/B for any complex semisimple Lie group G and Borel subgroup B. We should start with any irreducible representation of G and expect to resolve it by a complex of homogeneous line bundles linked by invariant operators. Specifically, if λ is a dominant integral weight for G, then $E(\lambda)$ is resolved by a complex whose structure is exactly the same as the Weyl group $W_{\mathbf{g}}$ of G with the line bundles in question obtained by the affine action of $W_{\mathbf{g}}$ on λ. In other words, we have

Theorem (8.3.1).   *If λ is a dominant integral weight for $\mathbf{g}$, then there is an exact resolution*

$$0 \to E(\lambda) \to \Delta^{\bullet}(\lambda)$$

where

$$\Delta^{p}(\lambda) = \bigoplus_{w \in W_{\mathbf{g}}, \ell(w)=p} \mathcal{O}_{\mathbf{b}}(w.\lambda)$$

and where the differential $\Delta^{p}(\lambda) \to \Delta^{p+1}(\lambda)$ is obtained by taking the direct sum of the operators

$$\mathcal{O}_{\mathbf{b}}(v.\lambda) \to \mathcal{O}_{\mathbf{b}}(w.\lambda)$$

where $w = \sigma v$ for some reflection σ (not necessarily simple).

In fact, Bernstein et al. [17] show much more than the general validity of this resolution in that they demonstrate that a non-zero invariant operator

$$\mathcal{O}_b(u.\lambda) \to \mathcal{O}_b(v.\lambda)$$

exists precisely when $u \preceq v$ in W_g in which case this operator is unique up to scale. Indeed, they use this fact, together with a combinatorial argument concerning the structure of W_g, to prove exactness of their resolution.

Alternatively, the resolution may be derived from the de Rham sequence as in our initial example. From this point of view, it is clear that

$$\Delta^1 = \bigoplus_{\sigma \in \Delta} \mathcal{O}_b(\sigma.0)$$

is a reasonable choice for the term following $\mathcal{O}$ in a resolution of C since it is dual to the smallest homogeneous subbundle $D \subseteq T$ with the property that $[D, D] = T$. The notion of *infinitesimal character* (see 3. on page 177) determines the remaining terms. The general case then follows by translation. Presumably, the machinery of Spencer [145] also generates the BGG resolution if we specify the composition $\mathcal{O} \xrightarrow{d} \Omega^1 \to \Delta^1$ as the first differential operator δ of the complex.

A typical BGG resolution (for the complete flag manifold of $SO(5, \mathbf{C})$ or $SP(2, \mathbf{C})$) is as follows.

8.4   The story for G/P

The discussion so far has been restricted to complete flag manifolds, i.e., of the form G/B for B a Borel subgroup of a semisimple G. However, there is a similar resolution on any generalized flag manifold, i.e., of the form G/P for P parabolic. The pattern on such a manifold follows the partially ordered set W^P as discussed in section 4.3. In other words, we have an analogue of theorem 8.3.1 above:

Theorem (8.4.1).   *If λ is a dominant integral weight for* **g**, *then there is an exact resolution*

$$0 \to E(\lambda) \to \Delta^\bullet(\lambda)$$

where

$$\Delta^P(\lambda) = \bigoplus_{w \in W^P, \ell(w) = p} \mathcal{O}_p(w.\lambda).$$

A typical example is the following resolution of $E(\overset{1}{\bullet}\!\!-\!\!\overset{0}{\bullet}\!\!-\!\!\overset{0}{\bullet})$:

$$\longrightarrow \overset{1\ \ 0\ \ 0}{\bullet\!-\!\times\!-\!\bullet} \longrightarrow \overset{2\ -2\ \ 1}{\bullet\!-\!\times\!-\!\bullet} \begin{array}{c} \nearrow \\ \\ \searrow \end{array} \begin{array}{c} \overset{3\ -3\ \ 0}{\bullet\!-\!\times\!-\!\bullet} \\ \oplus \\ \overset{0\ -4\ \ 3}{\bullet\!-\!\times\!-\!\bullet} \end{array} \begin{array}{c} \searrow \\ \\ \nearrow \end{array} \overset{1\ -5\ \ 2}{\bullet\!-\!\times\!-\!\bullet} \longrightarrow \overset{0\ -5\ \ 1}{\bullet\!-\!\times\!-\!\bullet} \longrightarrow 0.$$

Notice that this resolution is no longer composed of line bundles but, rather, of irreducible homogeneous vector bundles. This particular resolution has a well-known physical consequence [31,44] on $\bullet\!\!-\!\!\times\!\!-\!\!\bullet$ regarded as Minkowski space, namely that helicity $-\tfrac{3}{2}$ massless fields

$$\mathrm{ker}:\overset{0\ -4\ \ 3}{\bullet\!-\!\times\!-\!\bullet} \to \overset{1\ -5\ \ 2}{\bullet\!-\!\times\!-\!\bullet}$$

may be locally represented as potentials/gauge (see page 103)

$$\frac{\mathrm{ker}:\overset{2\ -2\ \ 1}{\bullet\!-\!\times\!-\!\bullet} \to \overset{3\ -3\ \ 0}{\bullet\!-\!\times\!-\!\bullet}}{\mathrm{im}:\overset{1\ \ 0\ \ 0}{\bullet\!-\!\times\!-\!\bullet} \to \overset{2\ -2\ \ 1}{\bullet\!-\!\times\!-\!\bullet}}.$$

The general case is proved in [66,110,131]. However, more in the geometric spirit of this book, these BGG resolutions on G/P may be derived from those on G/B as follows. Consider the natural projection

$$G/B \xrightarrow{\pi} G/P.$$

The fibres of this projection are themselves homogeneous as quotients

$$P/B = L/L \cap B$$

where L is the reductive part of P as in section 2.2. Thus, homogeneous cohomology along the fibres of π may be computed by BBW. In other words, any homogeneous line bundle $\mathcal{O}_{\mathbf{b}}(\lambda)$ on G/B, we can use BBW to compute its direct images $\pi_*^P \mathcal{O}_{\mathbf{b}}(\lambda)$. Specifically, the answer is as follows. Consider the affine action of the Weyl group $W_{\mathbf{p}}$ on the weight λ. If the orbit of λ under this action contains no weight dominant for P, then all the direct images are zero. Otherwise, λ is said to be *non-singular* for P and $W_{\mathbf{g}}.\lambda$ contains precisely one weight dominant for P. If $w.\lambda$ is this weight, then

$$\pi_*^{\ell(w)} \mathcal{O}_{\mathbf{b}}(\lambda) = \mathcal{O}_{\mathbf{p}}(w.\lambda_{\mathbf{p}})$$

and all other direct images vanish. Consider now a BGG resolution

$$0 \to E(\lambda) \to \Delta^\bullet$$

on G/B. Then $\pi_*^r E(\lambda) = \pi_*^r \Delta^\bullet(\lambda)$. But, $E(\lambda)$ is a constant sheaf on G/B, so its direct images are given by

$$\pi_*^r E(\lambda) = H^r(P/B, \mathbf{C}) \otimes E(\lambda)$$

and, from the cell structure of generalized flag manifolds discussed earlier in section 6.2,

$$\dim H^{2r}(P/B, \mathbf{C}) = \#\{w \in W_\mathbf{p} \text{ s.t. } \ell(w) = r\}.$$

On the other hand, $\pi_*^r \Delta^\bullet(\lambda)$ may be computed by the hyperdirect image spectral sequence

$$E_1^{p,q} = \pi_*^q \Delta^p(\lambda) \Longrightarrow \pi_*^{p+q} \Delta^\bullet(\lambda)$$

whose E_1-terms are given by BBW as above. For ease of discussion, suppose that λ is non-singular (for B and hence for P). Indeed, we could just do the case $\lambda = 0$ and use the Jantzen–Zuckerman translation to arrive at all other cases. Then recall (section 4.3) that $W^\mathbf{P}$ is characterized inside the Weyl group $W_\mathbf{g}$ of G as

$$W^\mathbf{P} = \{w \in W \text{ s.t. } w.\lambda \text{ is dominant for } P\}.$$

Thus, $E_1^{p,0}$ is a copy of what we are trying to prove is the BGG resolution of $E(\lambda)$. Indeed, since $W^\mathbf{P}$ consists of canonical representatives for the right cosets of $W_\mathbf{p}$, it follows from BBW that the E_1-level of the spectral sequence is made up of $W_\mathbf{p}$ copies of this proposed resolution emanating from the $E_1^{s,s}$-positions with multiplicity $= \#\{w \in W_\mathbf{p} \text{ s.t. } \ell(w) = s\}$. The BGG resolution on ⤨⤨ as above gives a typical example under projection to ⦁⤨.

In general, if these were resolutions of $E(\lambda)$, then the E_2-level would consist of $E(\lambda)$'s scattered along the diagonal $E_2^{s,s}$ with appropriate multiplicity. The spectral sequence would therefore collapse and give the correct answer for the direct images $\pi_*^r E(\lambda)$. Conversely, it is clear that this is the only way that the spectral sequence is going to converge to the right answer.

Notice that, as for complete flag manifolds, the BGG resolution of $\mathcal{O}$ on any generalized flag manifold is closely related to the de Rham resolution. In particular, the bundle Δ^1 has a similar geometrical interpretation as dual to the smallest homogeneous subbundle $D \subseteq T$ with the property that $[D, D] = T$. As before, the BGG resolution may, alternatively, be derived from the de Rham resolution (cf. [51]) and if the tangent bundle happens to be irreducible (see table 3.2) then the two resolutions will coincide. This happens, for example, in the case of Minkowski space $\mathbf{M}$ as the homogeneous manifold $\bullet\!\!-\!\!\times\!\!-\!\!\bullet$ where the BGG resolution

coincides with the de Rham resolution

the two-forms being split $\Omega^2 = \Omega^2_+ \oplus \Omega^2_-$ into *self-dual* and *anti-self-dual* types by virtue of the conformal structure on $\mathbf{M}$ (see example 2.3.2 and remark 3.2.5).

In fact, for any complex quadric the tangent bundle is an irreducible homogeneous bundle and the de Rham and BGG resolutions agree—this gives a simple method for computing the form bundles on a quadric as homogeneous bundles; the reader may wish to use this observation to verify example 3.2.4.

8.5 An algorithm for computation

We can use the method for determining $W^{\mathbf{p}}$ given in section 4.3 to calculate a BGG resolution. This really is the key computation in constructing a Penrose transform so it is worth illustrating the algorithm in detail.

Example (8.5.1).   To determine a BGG resolution on ambitwistor space $\times\!\!-\!\!\bullet\!\!-\!\!\times$ we should first allow the Weyl group $W_{\mathbf{g}}$ to act on the weight $\begin{smallmatrix} 1 & 0 & 1 \\ \bullet\!-\!\bullet\!-\!\bullet \end{smallmatrix}$, obtaining the result

In this diagram only the simple reflections are shown. To use this to create the BGG resolution of $\overset{0\ \ \ 1\ \ \ 2}{\bullet\!-\!\bullet\!-\!\bullet}$, for example, consider the composition

$$\overset{1\quad 0\quad 1}{\bullet\!-\!\bullet\!-\!\bullet}\ \overset{\sigma}{\mapsto}\ \overset{-1\quad 1\quad 1}{\bullet\!-\!\bullet\!-\!\bullet}\ \overset{\tau}{\mapsto}\ \overset{0\quad -1\quad 2}{\bullet\!-\!\bullet\!-\!\bullet}.$$

Inverting this composition and computing the affine action on $\overset{0\ \ \ 1\ \ \ 2}{\bullet\!-\!\bullet\!-\!\bullet}$ gives

$$\overset{0\quad 1\quad 2}{\bullet\!-\!\bullet\!-\!\bullet}\ \overset{\tau}{\mapsto}\ \overset{2\quad -3\quad 4}{\bullet\!-\!\bullet\!-\!\bullet}\ \overset{\sigma}{\mapsto}\ \overset{-4\quad 0\quad 4}{\bullet\!-\!\bullet\!-\!\bullet}.$$

This is therefore the appropriate weight for the corresponding position in the resolution. Adding in the non-simple reflections gives

as a resolution of $\overset{0\ \ \ 1\ \ \ 2}{\bullet\!-\!\bullet\!-\!\bullet}$ on $\times\!-\!\bullet\!-\!\times$ using irreducible homogeneous bundles and invariant holomorphic differential operators between them.

Remark (8.5.2). The algorithm used to compute a BGG resolution may be applied to a singular (but dominant) weight. The result is an exact sequence, still, which is no longer a resolution of a finite dimensional representation. For example, the sequences

and

are exact on $\bullet\!-\!\times\!-\!\bullet$.

8.6 Non-standard morphisms

In their article [17], not only did Bernstein et al. prove the exactness of their resolution on G/B but also they classified all the invariant differential operators between homogeneous line bundles on G/B. The result is that, up to scale, they are precisely those which arise in the resolution together with compositions thereof (as in theorem 11.2.2). We might expect that the invariant operators on G/P are similarly classified. However, this is not the case. In other words, there are operators on G/P which do not arise as direct images of operators on G/B under the natural projection $G/B \to G/P$. These operators are called *non-standard* [19,20,110]. There are non-standard operators in the singular case too and in fact the simplest example is an extra operator in the previous diagram

$$\square: \overset{0 \quad\, -1 \quad 0}{\bullet\!\!-\!\!\underset{\times}{\bullet}\!\!-\!\!\bullet} \longrightarrow \overset{0 \quad\, -3 \quad 0}{\bullet\!\!-\!\!\underset{\times}{\bullet}\!\!-\!\!\bullet}$$

which is the *wave operator* or *Laplacian*. More generally, there is an extra operator

$$\overset{p \quad q \quad r}{\bullet\!\!-\!\!\underset{\times}{\bullet}\!\!-\!\!\bullet} \longrightarrow \underset{\text{-p-q-r-4}}{\overset{r \quad\quad\ p}{\bullet\!\!-\!\!\underset{\times}{\bullet}\!\!-\!\!\bullet}}$$

and, together with those which arise from the BGG resolution, this forms a complete list of $SL(4, \mathbf{C})$ invariant operators (see [53] for more details). Although these operators do not arise as direct images of operators on $\times\!\!-\!\!\times\!\!-\!\!\times$, they arise naturally from the Penrose transform (the wave operator is explained on page 107 and a more exotic example given in section 9.8). The ingredients for this transform, however, are all standard, i.e., the construction relies only on the BGG resolution. We may hope, therefore, that *all* invariant operators arise from the Penrose transform. This would be a very useful phenomenon, the general invariant operator being unknown at present. We shall take up this question again below and show that the Penrose transform *does* generate all conformally invariant differential operators.

8.7   Relative BGG resolutions

The final variation which we shall need concerns *relative* BGG resolutions. These are resolutions along the fibres of a projection $G/Q \overset{\eta}{\to} G/R$ for the case of two standard parabolic subgroups $Q \subseteq R \subseteq G$. Since the fibres are themselves homogeneous, we can construct a BGG resolution on each. Thus, given an irreducible homogeneous bundle E on G/R, we can restrict to each point $x \in G/R$ and resolve along the fibre $\eta^{-1}(x)$. It is then easy to see that the resulting resolutions vary in a homogeneous manner as we vary the fibre. In other words, the resolution is by invariant differential operators acting between irreducible homogeneous bundles on G/Q. The ordinary BGG resolutions on G/Q correspond to the special case when $R = G$ so that η is just collapsing to a point. In general, the relative BGG construction resolves the topological inverse image sheaf $\eta^{-1}\mathcal{O}(E)$. An alternative resolution is given by the relative holomorphic de Rham sequence and until recently [44,50] it was this resolution which was used to describe the Penrose transform. The use of the BGG resolution and, indeed, the Jantzen–Zuckerman functor for its construction is implicit in [51] for the special case of Minkowski space where to use the de Rham sequence for a general homogeneous bundle is already unmanageable. More generally, the BGG resolution is absolutely crucial in order to avoid the difficulties which inevitably arise with reducible tangent bundles. The algorithm for computing these relative resolutions is a straightforward generalization of the previous discussion.

To compute a relative BGG resolution, simply carry out the algorithm as in example 8.5.1 but use $W_{\mathbf{r}}^{\mathbf{q}}$, the relative Hasse diagram of the fibration (see section 4.4).

Example (8.7.1).   Consider the fibration

To compute the relative BGG resolution for this fibration the first step is to compute the orbit of $\overset{0\ \ 1\ \ 0}{\bullet\!-\!\!-\!\bullet\!-\!\!-\!\bullet}$ under $W_{\mathbf{r}}$ (which is generated by σ_{α_2} and σ_{α_3}) as

$$\overset{0\quad 1\quad 0}{\bullet\!-\!\bullet\!-\!\bullet} \;\rightarrow\; \overset{1\quad -1\quad 1}{\bullet\!-\!\bullet\!-\!\bullet} \;\rightarrow\; \overset{1\quad 0\quad -1}{\bullet\!-\!\bullet\!-\!\bullet}$$

which gives

$$W_{\mathbf{r}}^{\mathbf{q}} = \{id, \sigma_{\alpha_2}, \sigma_{\alpha_2}\sigma_{\alpha_3}\}$$

from which we deduce the resolution

$$0 \rightarrow \eta^{-1}\overset{k\quad 0\quad 0}{\underset{\times\!-\!\bullet\!-\!\bullet}{}} \rightarrow \overset{k\quad 0\quad 0}{\underset{\times\!-\!\times\!-\!\bullet}{}} \rightarrow \overset{k+1\ -2\ \ 1}{\underset{\times\!-\!\times\!-\!\bullet}{}} \rightarrow \overset{k+2\ -3\ \ 0}{\underset{\times\!-\!\times\!-\!\bullet}{}} \rightarrow 0.$$

This is the start of the classical Penrose transform and will be used in section 9.2 below.

9

THE PENROSE TRANSFORM IN PRACTICE

This chapter is devoted to the calculation of several examples of the Penrose transform on generalized flag manifolds. We shall illustrate the computations involved in some detail—it is our intention that this book should be useful as a handbook of techniques. Of course, an endless repetition of an unvarying theme is tiresome, so we will use the opportunity to illustrate the large variety of applications of the transform. Some of these are as follows.

Firstly, and historically, the Penrose transform gives an algebraic construction of solutions to a wide class of physically important differential equations, as pointed out in the introductory chapters. It provides a means of writing down these solutions quite explicitly in closed integral form. More significantly, it gives a definite structure to the space of all solutions of the equation by recognizing it as a sheaf cohomology group on an associated complex manifold. This gives us a chance to directly address problems involving the total structure of the solution space. For example, we may hope to manipulate the solution space geometrically to *quantize* the classical system associated to the original differential equations. In doing this it seems essential that the space of solutions should admit a *unitary structure*, invariant under a suitable group action.

This gives a second consequence of the transform. For in a form often called the *twistor transform*, it provides a direct method of constructing an invariant inner product on many of the cohomology groups (without having to resort to the use of L^2-cohomology). It turns out not to be too difficult to show the definiteness of this inner product in good cases. Not all cohomology groups will admit this structure, since, as we shall see, their K-finite vectors form a subquotient of a dual Verma module. Not all such subquotients are unitarizable [61]. We investigate the twistor transform and take up the unitary question in chapter 10.

Thirdly, as has already been mentioned, the differential operators which arise in the Penrose transform on a generalized flag manifold are invariant under left translations. They correspond to homomorphisms of generalized Verma modules [53,110]. Actually, these operators arise in two stages: firstly as differentials in the Bernstein–Gelfand–Gelfand resolution (the "pull-back" part of the construction) and secondly as the derived direct images of these differentials (in the hypercohomology part of the transform). Now, taking direct images of the Bernstein–Gelfand–Gelfand differentials corresponds to pushing down a (generalized) Verma module homomorphism from a module induced using a small parabolic to one induced using a larger parabolic. This is a standard procedure; the result is always defined, but may be zero. In this eventuality, surprisingly, there may still exist a non-zero *non-standard* homomorphism for the larger parabolic [110]. The Penrose transform handles this situation easily, producing non-standard homomorphisms directly. The reason lies in the fact that the transform considers derived forms of the homomorphisms. The occurrence of non-standard homomorphisms is exactly the occurrence of irreducible subquotients of Verma modules with higher multiplicity than one. So the transform gives structural information on Verma modules.

Fourthly, the Penrose transform has several applications in differential geometry. More complete information is given in [90,93,123,128,155] and we briefly study only two. The cohomology of tangent bundles on generalized flag manifolds (or subsets thereof, more correctly) relates to the *holomorphic deformation theory* of such manifolds [101]. If the tangent bundle is not reduced then the cohomology of (sub)quotient irreducible sheaves corresponds to deformations preserving the part of the composition series structure of the tangent bundle. The simplest case of this is when the manifold admits a *contact structure* which must be preserved under deformation. Spaces of null geodesics in a conformal manifold are precisely of this form [105] and the Penrose transform then realizes deformations of these spaces which *preserve* the contact structure as deformation of the conformal structure of the manifold in which the geodesics live.

The chapter is divided as follows: in section 9.2, we review the historical examples of the transform originally due to Roger Penrose [119]. Thus we compute the cohomology of powers of the tautological bundle on $\mathbf{CP}^3$ restricted to the neighbourhood of a line and identify these with solutions of the *zero-rest-mass* field equations on an affine region of Minkowski space. This prototype has many of the features of the more general setting. We also consider the *twistor transform*. Sections 9.3 and 9.4 consider more general homogeneous bundles over $\mathbf{CP}^3$ such as the holomorphic forms and the tangent bundle. Another useful version of the transform in the physically relevant case of four dimensions is that for "ambitwistors" and this is the subject of section 9.5. We point out how infinitesimal character explains the calculations involved in the extension problems on ambitwistor space.

Next we consider higher dimensional analogues of the Penrose transform. From the physical point of view there are two possibilities—we may generalize the *conformal* structure of four dimensional Minkowski space or (what amounts to the same structure in four dimensions) the fact that the tangent bundle occurs as a tensor product of two more basic bundles. The first possibility is considered in section 9.6, where we discuss quadrics of arbitrary dimension. On the twistor spaces which we introduced for these spaces in section 2.3, there is a natural line bundle which generalizes the tautological bundle. Taking cohomology of its powers on a suitable region of the twistor space leads again to solutions of higher analogues of the zero rest mass field equations. The second possibility is the subject of section 9.7. For this, G/P is the Grassmannian $\mathbf{Gr}_p(\mathbf{C}^{p+q})$ and the twistor spaces are the adjacent Grassmannians $\mathbf{Gr}_{p\pm1}(\mathbf{C}^{p+q})$. Grassmannians occur because their tangent bundles are naturally a product of bundles—if S' denotes the tautological bundle on a Grassmannian and Q its quotient bundle, then its tangent bundle is $S'^* \otimes Q$. In many ways, the resulting twistor theory is a closer analogue of the four dimensional case than the conformal generalization.

The machinery that has been developed in this book easily deals with many variations on the Penrose transform. This is illustrated in section 9.8 with an example homogeneous under the complex Lie group E_6. There is a "non-Abelian" version of the Penrose transform known as the Ward correspondence [153]. This is the subject of section 9.9.

9.1   The homogeneous Penrose transform

Fix G and standard parabolics P, R so that $Q = P \cap R$ is parabolic. The basic correspondence to study is the following:

$$
\begin{array}{ccc}
 & G/Q & \\
{}^{\eta}\swarrow & & \searrow^{\tau} \\
G/R & & G/P.
\end{array}
$$

Let $X \subset G/P$ be open and put $Y = \tau^{-1}X$, and $Z = \eta Y$; we assume for the most part that X is affine or Stein, and certainly contractible, so that the pull-back stage of the Penrose transform is trivial:

$$
H^i(Z, \mathcal{F}) \cong H^i(Y, \eta^{-1}\mathcal{F}).
$$

Take $\mathcal{F} = \mathcal{O}_{\mathbf{r}}(\lambda)$ restricted to Y. Then on Y there is a Bernstein–Gelfand–Gelfand resolution:

$$0 \to \eta^{-1}\mathcal{O}_{\mathbf{r}}(\lambda) \to \Delta_\eta^\bullet(\lambda)$$

where

$$\Delta_\eta^p(\lambda) = \bigoplus_{w \in W_{\mathbf{r}}^{\mathbf{q}};\, \ell(w)=p} \mathcal{O}_{\mathbf{q}}(w.\lambda).$$

Assuming that X is Stein or affine, the Leray spectral sequence collapses to give a series of isomorphisms:

$$H^q(Y, \Delta_\eta^\bullet(\lambda)) \cong \Gamma(X, \tau_*^q \Delta_\eta^\bullet(\lambda))$$

where τ_*^q denotes the qth direct image and is computed by means of the Bott–Borel–Weil theorem. Lastly, we may use the *hypercohomology spectral sequence*

$$E_1^{p,q} = H^p(Y, \Delta_\eta^q(\lambda)) \implies H^{p+q}(Y, \eta^{-1}\mathcal{O}_{\mathbf{r}}(\lambda))$$

or, in other words,

$$E_1^{p,q} = \Gamma(X, \tau_*^q \Delta_\eta^p(\lambda)) \implies H^{p+q}(Z, \mathcal{O}_{\mathbf{r}}(\lambda))$$

as in theorem 7.3.1 to complete the transform. The differential equations on X are induced by the differentials of this spectral sequence which are, in turn, derived from those of the BGG resolution.

This machine is really best understood when one uses it! So, rather than try to elaborate this brief recipe we shall now give many examples of its practical calculation. We strongly recommend that the reader don her/his overalls, fetch lots of fresh paper and pencils, and calculate along with us.

9.2   The real thing

In this section, we illustrate the machinery of the Penrose transform by rederiving the original results of Penrose [44,119,124,125]. He considered the holomorphic cohomology of powers of the tautological sheaf on $\mathbf{CP}^3$ restricted to a neighbourhood of a line and showed that it is isomorphic with the solution spaces of the so-called *zero-rest-mass* field equations on the corresponding region of Minkowski space. For this, $G = \mathrm{SL}(4, \mathbf{C})$ and the basic double fibration is

$$\mathbf{CP}^3 = \times\!\!-\!\!\bullet\!\!-\!\!\bullet \qquad\qquad \bullet\!\!-\!\!\times\!\!-\!\!\bullet = \mathbf{Gr}_2(\mathbf{C}^4) = \mathbf{M}.$$

As the image variety of the transform we choose a Stein subset of Minkowski space. From the algebraic point of view, this could be the open affine cell surrounding a base point (the identity coset, say). That is, $\mathbf{M}^I = \mathbf{CS}^4 \setminus \{\text{light cone at infinity}\}$. This is the set X of the previous section. The Y is the set $\mathbf{F}^I = \tau^{-1}\mathbf{M}^I$ and the corresponding subset of the twistor space $\mathbf{P}$ is easily identified as $\mathbf{P}^I = \mathbf{CP}^3 \setminus L$, where L is the line whose points correspond to totally null α-planes contained in the light cone at infinity. Thus we are considering the following double fibration:

In what follows, we shall want to give explicit formulae (in the abstract index notation of [127]) for the differential operators produced by the transform. To this end, it is convenient to fix a flat metric on $\mathbf{M}^I$ and have available its Levi–Civita connection ∇_a, acting on sections of all homogeneous sheaves. Such a metric is easy to construct. The open affine cell $\mathbf{M}^I$ is the orbit of a distinguished parabolic subgroup P of $\mathrm{SL}(4,\mathbf{C})$ (which is a $4-1$ covering of the Poincaré group): the stabilizer of a point of $\mathbf{M}^I$ is a maximal reductive subgroup isomorphic to $\mathrm{Spin}(4,\mathbf{C})$. P stabilizes an extremal weight space in $\overset{0}{\bullet}\!\!-\!\!\overset{1}{\bullet}\!\!-\!\!\overset{0}{\bullet}$ (the lowest, say), so fixing a skew twistor $I_{[\alpha\beta]}$ up to scale. This gives an element of $\Gamma(\mathbf{CS}^4, \overset{0}{\bullet}\!\!-\!\!\overset{1}{\underset{\times}{\bullet}}\!\!-\!\!\overset{0}{\bullet})$ whose zero variety is the "light cone at infinity". Its square, in $\Gamma(\mathbf{CS}^4, \overset{0}{\bullet}\!\!-\!\!\overset{2}{\underset{\times}{\bullet}}\!\!-\!\!\overset{0}{\bullet})$, is the desired flat metric. The associated (affine) Levi–Civita connection corresponds to the Maurer–Cartan form on P, and parallel transport corresponds to left translation by P. An irreducible homogeneous sheaf on $\mathbf{CS}^4$ evidently gives such a sheaf on $\mathbf{M}^I$; note, however, that restrictions of irreducible $\mathrm{SL}(4,\mathbf{C})$ representations to $\mathrm{Spin}(4,\mathbf{C})$ split into direct sums of irreducibles, so that on $\mathbf{M}^I$ exact sequences such as

$$0 \to \overset{1}{\bullet}\!\!-\!\!\overset{-1}{\underset{\times}{\bullet}}\!\!-\!\!\overset{0}{\bullet} \to \overset{0}{\bullet}\!\!-\!\!\overset{0}{\bullet}\!\!-\!\!\overset{1}{\bullet} \to \overset{0}{\bullet}\!\!-\!\!\overset{0}{\underset{\times}{\bullet}}\!\!-\!\!\overset{1}{\bullet} \to 0$$

split canonically over $\mathcal{O}_{\mathbf{M}^I}$ once a base point has been fixed. We shall use this in our discussion of the wave operator (or complex Laplacian) below.

Let $\mathcal{O}(k) = \overset{k}{\underset{\times}{\bullet}}\!\!-\!\!\overset{0}{\bullet}\!\!-\!\!\overset{0}{\bullet}$ be the kth power of the tautological sheaf on $\mathbf{CP}^3$, restricted to subsets, where appropriate.

We distinguish three cases according as $k+2$ is negative, zero, or positive. In physics, these cases correspond to solving the *positive helicity* zero rest mass (ZRM) field equations, the conformally invariant wave equation, and providing *potentials modulo gauge* for solutions of the *negative helicity* zero rest mass equations. From the point of view of the representation

theorist, they correspond to varying degrees of singularity for the weight
$\overset{k}{\bullet}\!\!-\!\!\overset{0}{\bullet}\!\!-\!\!\overset{0}{\bullet}$. Not surprisingly, the details of the transform vary markedly in
each case. We take the easiest first.

Positive helicity ZRM *fields:* $k + 2 < 0$

Because the fibres over $\mathbf{P}^I$ are topologically trivial, the cohomology
$H^*(\mathbf{P}^I, \mathcal{O}(k))$ is identified with $H^*(\mathbf{F}^I, \eta^{-1}\mathcal{O}(k))$. We compute this. The
first step is to calculate the Bernstein–Gelfand–Gelfand resolution of
$\eta^{-1}\mathcal{O}(k)$ on the fibres of η. This is

$$0 \to \eta^{-1}\mathcal{O}(k) \to \overset{k\ \ \ \ 0\ \ \ \ 0}{\times\!\!-\!\!\times\!\!-\!\!\bullet} \to \overset{k+1\ -2\ \ \ 1}{\times\!\!-\!\!\times\!\!-\!\!\bullet} \to \overset{k+2\ -3\ \ \ 0}{\times\!\!-\!\!\times\!\!-\!\!\bullet} \to 0.$$

Now we take direct images of this resolution, $\mathbf{\Delta}_\eta^\bullet$. Using the (relative) Bott–
Borel–Weil theorem of chapter 5 we obtain:

$$\tau_*^1\!\left(\overset{k\ \ \ \ 0\ \ \ \ 0}{\times\!\!-\!\!\times\!\!-\!\!\bullet}\right) = \overset{-k\text{-}2\ \ k+1\ \ 0}{\bullet\!\!-\!\!\times\!\!-\!\!\bullet} = \mathcal{O}_{\underbrace{(A'B'C'\ldots D')}_{-k-2}}[-1]$$

$$\tau_*^1\!\left(\overset{k+1\ -2\ \ \ 1}{\times\!\!-\!\!\times\!\!-\!\!\bullet}\right) = \overset{-k\text{-}3\ \ k\ \ \ 1}{\bullet\!\!-\!\!\times\!\!-\!\!\bullet} = \mathcal{O}_{B\underbrace{(B'C'\ldots D')}_{-k-3}}[-2]$$

$$\tau_*^1\!\left(\overset{k+2\ -3\ \ \ 0}{\times\!\!-\!\!\times\!\!-\!\!\bullet}\right) = \begin{cases} \overset{k+1\ -2\ \ \ 1}{\bullet\!\!-\!\!\times\!\!-\!\!\bullet} \\[4pt] 0 \end{cases} = \begin{cases} \mathcal{O}_{\underbrace{(C'\ldots D')}_{-k-4}}[-4] & \text{if } k \le -4 \\[10pt] 0 & \text{if } k = -3. \end{cases}$$

Because $k + 2 < 0$, only *first* direct images occur. (We have indicated, also,
the abstract index notation for the direct images on Minkowski space [127].)
Notice that if $k + 3 < 0$ then the weight $\overset{k+2\ -3\ \ \ 0}{\bullet\!\!-\!\!\bullet\!\!-\!\!\bullet}$ is non-singular: thus, all
resolving sheaves have exactly one non-zero direct image. The case $k = -3$
is singular, a fact mirrored in the vanishing of all direct images of the last
resolvent. This only makes things easier, so let us treat the remainder of
the transform for it, first. The next step is to compute the cohomology
of each Δ_η^i on the intermediate space in terms of the cohomology of their
direct images on the target subset of Minkowski space. Because the image
variety of the transform is affine, the Leray spectral sequence collapses to
a series of isomorphisms:

$$H^j(\mathbf{F}^I, \Delta_\eta^i) \cong \Gamma(\mathbf{M}^I, \tau_*^j \Delta_\eta^i).$$

Substitute these in the hypercohomology spectral sequence to compute the
cohomology $H^*(\mathbf{P}^I, \mathcal{O}(-3))$. The $E_1^{p,q}$ term is as follows (for notational

brevity, the sections of a sheaf over $\mathbf{M}^I$ are denoted by the sheaf itself in this and subsequent spectral sequences):

$$E_1^{p,q} : \quad \begin{array}{ccc} 0 & 0 & 0 \\[4pt] \overset{1 \;\; -2 \;\; 0}{\bullet\!\!-\!\!\times\!\!-\!\!\bullet} \;\; \overset{0 \;\; -3 \;\; 1}{\bullet\!\!-\!\!\times\!\!-\!\!\bullet} & & 0 \\[4pt] 0 & 0 & 0 \end{array}$$

Deriving once, obtain the $E_2^{p,q}$ term:

$$E_2^{p,q} : \quad \begin{array}{ccc} 0 & 0 & 0 \\[4pt] \text{ker:} \;\; \overset{1 \;\; -2 \;\; 0}{\bullet\!\!-\!\!\times\!\!-\!\!\bullet} \to \overset{0 \;\; -3 \;\; 1}{\bullet\!\!-\!\!\times\!\!-\!\!\bullet} & \text{coker:} \;\; \overset{1 \;\; -2 \;\; 0}{\bullet\!\!-\!\!\times\!\!-\!\!\bullet} \to \overset{0 \;\; -3 \;\; 1}{\bullet\!\!-\!\!\times\!\!-\!\!\bullet} & 0 \\[4pt] 0 & 0 & 0 \end{array}$$

All differentials are now zero, so $E_2^{p,q} = E_\infty^{p,q}$ from which we deduce that

$$H^1(\mathbf{P}^I, \mathcal{O}(-3)) \cong \ker \Gamma(\mathbf{M}^I, \overset{1 \;\; -2 \;\; 0}{\bullet\!\!-\!\!\times\!\!-\!\!\bullet}) \overset{d}{\to} \Gamma(\mathbf{M}^I, \overset{0 \;\; -3 \;\; 1}{\bullet\!\!-\!\!\times\!\!-\!\!\bullet})$$

$$\cong \ker \Gamma(\mathbf{M}^I, \mathcal{O}_{A'}[-1]) \to \Gamma(\mathbf{M}^I, \mathcal{O}_A[-3]).$$

It remains to identify the differential operator d in the right-hand terms of this isomorphism. To do this, we identify its symbol, since this determines such an invariant differential operator [53]. Recall that $\overset{1 \;\; -2 \;\; 1}{\bullet\!\!-\!\!\times\!\!-\!\!\bullet} = \mathcal{O}_{AA'}$ is the holomorphic tangent bundle of Minkowski space. The only possible symbols are proportional to the projection

$$\overset{0 \;\; -3 \;\; 1}{\bullet\!\!-\!\!\times\!\!-\!\!\bullet} \otimes \overset{1 \;\; -2 \;\; 1}{\bullet\!\!-\!\!\times\!\!-\!\!\bullet} \cong \overset{2 \;\; -4 \;\; 1}{\bullet\!\!-\!\!\times\!\!-\!\!\bullet} \oplus \overset{0 \;\; -3 \;\; 1}{\bullet\!\!-\!\!\times\!\!-\!\!\bullet} \to \overset{0 \;\; -3 \;\; 1}{\bullet\!\!-\!\!\times\!\!-\!\!\bullet}$$

or, in abstract index notation,

$$\mathcal{O}_{A'}[-1] \otimes \mathcal{O}_{AA'} \cong \mathcal{O}_{A(A'B')}[-1] \oplus \mathcal{O}_A[-3] \to \mathcal{O}_A[-3].$$

This is the symbol of the *Dirac operator* [38] which is given by

$$\varphi_{A'} \to \nabla_A^{A'} \varphi_{A'}.$$

In a moment we shall argue that d is non-zero and so the first cohomology of $\mathcal{O}(-3)$ on twistor space has been identified with the kernel of this operator on Minkowski space; that is with (classical) massless right-handed neutrinos.

Remark (9.2.1). It is easy to see that $\mathbf{P}^I$ can be covered by two Stein sets: write L as the intersection of two planes and consider the complements of these two planes. Thus, by the Mayer–Vietoris sequence, $H^p(\mathbf{P}^I, \mathcal{F})$ vanishes for any coherent sheaf $\mathcal{F}$ when $p > 1$. So, necessarily,

$$\Gamma(\mathbf{M}^I, \overset{1\ \ \ -2\ \ \ 0}{\bullet\!\!-\!\!\times\!\!-\!\!\bullet}) \to \Gamma(\mathbf{M}^I, \overset{0\ \ \ -3\ \ \ 1}{\bullet\!\!-\!\!\times\!\!-\!\!\bullet})$$

is surjective and, in particular, non-zero. This observation is very useful: it leads to the existence of many non-standard homomorphisms of Verma modules, as we shall see in section 9.8 and chapter 11.

Now suppose $k + 3 < 0$. The $E_1^{p,q}$ term of the hypercohomology spectral sequence is as follows:

$$E_1^{p,q} : \begin{array}{|ccc}
0 & 0 & 0 \\
\overset{-k-2\ \ k+1\ \ 0}{\bullet\!\!-\!\!\times\!\!-\!\!\bullet} & \overset{-k-3\ \ k\ \ 1}{\bullet\!\!-\!\!\times\!\!-\!\!\bullet} & \overset{-k-4\ \ k\ \ 0}{\bullet\!\!-\!\!\times\!\!-\!\!\bullet} \\
0 & 0 & 0
\end{array}$$

Now observe the following.

Remark (9.2.2). To continue remark 9.2.1, observe that the vanishing of second cohomology on $\mathbf{P}^I$ makes the second row of $E_1^{p,q}$ exact, except at $E_1^{0,1}$; indeed the row resolves the cohomology at this term.

This remark implies that

$$E_2^{p,q} = E_\infty^{p,q} : \begin{array}{|ccc}
0 & 0 & 0 \\
\text{ker:}\ \overset{-k-2\ \ k+1\ \ 0}{\bullet\!\!-\!\!\times\!\!-\!\!\bullet} \to \overset{-k-3\ \ k\ \ 1}{\bullet\!\!-\!\!\times\!\!-\!\!\bullet} & 0 & 0 \\
0 & 0 & 0
\end{array}$$

and again the spectral sequence converges after just one derivation. We deduce that

$$H^1(\mathbf{P}^I, \mathcal{O}(k)) \cong \ker \Gamma(\mathbf{M}^I, \overset{-k-2\ \ k+1\ \ 0}{\bullet\!\!-\!\!\times\!\!-\!\!\bullet}) \to \Gamma(\mathbf{M}^I, \overset{-k-3\ \ k\ \ 1}{\bullet\!\!-\!\!\times\!\!-\!\!\bullet})$$

$$\cong \ker \Gamma(\mathbf{M}^I, \mathcal{O}_{\underbrace{(A'B'\ldots D')}_{-k-2\ \text{terms}}}[-1]) \to \Gamma(\mathbf{M}^I, \mathcal{O}_{\underbrace{A(B'C'\ldots D')}_{-k-3\ \text{terms}}}[-3]). \quad (14)$$

The symbol of the differential operator is the projection

$$\overset{-k-2\ \ k+1\ \ 0}{\bullet\!\!-\!\!\times\!\!-\!\!\bullet} \otimes \overset{1\ \ \ -2\ \ \ 1}{\bullet\!\!-\!\!\times\!\!-\!\!\bullet} \cong \overset{-k-1\ \ k-1\ \ 1}{\bullet\!\!-\!\!\times\!\!-\!\!\bullet} \oplus \overset{-k-3\ \ k\ \ 1}{\bullet\!\!-\!\!\times\!\!-\!\!\bullet} \to \overset{-k-3\ \ k\ \ 1}{\bullet\!\!-\!\!\times\!\!-\!\!\bullet}$$

which is the symbol of the Dirac–Weyl operator on totally symmetric spinors of higher helicity:

$$\varphi_{A'B'\ldots C'} \to \nabla^{A'}_A \varphi_{A'B'\ldots C'}.$$

So, the first cohomology of $\mathcal{O}(k)$ $(k < -2)$ is identified with the kernel of this operator; in other words with solutions of the zero rest mass field equations of helicity $\frac{1}{2}(-k-2)$.

Remark (9.2.3).   The transform has been applied with an open affine subvariety of Minkowski space as a target space; we took this to be the largest left orbit $\mathbf{M}^I$ of the Borel subgroup B. It is just as interesting to take as target space an open orbit of a non-compact real form of SL(4, $\mathbf{C}$), such as SU(2, 2). As observed in chapter 6, there are two open orbits, $\mathbf{M}^+$ and $\mathbf{M}^-$, consisting of two-planes in $\mathbf{C}^4$ on which the restriction of the SU(2, 2)-preserved Hermitian form is positive and negative definite, respectively. The corresponding subspaces $\mathbf{P}^+$ and $\mathbf{P}^-$ of twistor space $\mathbf{CP}^3$ consist of those lines on which, again, the Hermitian form is positive or negative definite and the isomorphisms (14) hold with $\mathbf{M}^I$ and $\mathbf{P}^I$ replaced by $\mathbf{M}^-$ and $\mathbf{P}^+$.

Indeed, we may go even further and compute the transform on the *closures* of these regions, $\overline{\mathbf{M}}^+$ and $\overline{\mathbf{P}}^+$. Although $\overline{\mathbf{M}}^+$ is *not* a Stein manifold, it has arbitrarily small Stein neighbourhoods in $\mathbf{M}$, so the Leray spectral sequence degenerates in the same way. The resulting first cohomology group is naturally somewhat smaller (cf. page 167). Now $\overline{\mathbf{M}}^+$ has as Šilov boundary the unique closed orbit of SU(2, 2) which is just the set of *real* points of Minkowski space (relative to the conjugation defining SU(2, 2)). This is the conformal compactification of the real physical Minkowski space, and so this cohomology group is identified with real analytic solutions of the zero rest mass field equations on *real* Minkowski space which extend to the "forward tube" $\mathbf{M}^+$. Such fields are called *positive frequency* fields [146] and are of fundamental importance in physics.

The usual path followed in *quantizing* free zero rest mass fields begins by equipping the space of all such with a *unitary* structure. We will see in chapter 10 how this can be done using the *twistor transform*—that is, a double Penrose transform. It is possible to generalize this construction considerably to recover all discrete series representations for real reductive groups [41].

Remark (9.2.4).   Although the Stein property of $\mathbf{M}^+$ and $\mathbf{M}^I$ is a convenient one causing the Leray spectral sequence as in section 7.2 to collapse, in this case, since $\tau^0_* \Delta^\bullet_\eta = 0$, the spectral sequence of theorem 7.3.1 always has $E_1^{p,0} = 0$ and we continue to obtain the identification of positive helicity massless fields on $U \subseteq \mathbf{M}$ with the cohomology $H^1(Z, \mathcal{O}(k))$ provided that the argument of section 7.1 is still valid. In other words, the *analytical* properties of U are irrelevant whereas the *topological* restriction is that all α-planes intersect U in connected and simply connected regions.

Remark (9.2.5).   Another obvious variant is to compute the transform within the *algebraic* category. Since this naturally includes in the holomorphic category, we obtain a distinguished subset

$$H^1(\mathbf{P}^I, \mathcal{O}_{\mathrm{alg}}(k)) \hookrightarrow H^1(\mathbf{P}^I, \mathcal{O}(k))$$

where the subscript "alg" indicates the algebraic sheaf. The corresponding solutions of the field equations are polynomials on $\mathbf{M}^I$. These are often called *elementary states* [43,54,120].

We shall study elementary states in greater detail below where we will see that they can be identified with global sections of local cohomology sheaves; we will see that $H^1(\mathbf{P}^I, \mathcal{O}_{\mathrm{alg}}(k))$ is actually an irreducible lowest weight modules over $\mathcal{U}(\mathbf{sl}(4, \mathbf{C}))$ which shows that the mapping

$$H^1(\mathbf{P}^I, \mathcal{O}_{\mathrm{alg}}(k)) \rightarrow H^1(\overline{\mathbf{P}}^+, \mathcal{O}(k))$$

is an injection. The special significance of the elementary states is that they are *dense* in the latter group, when it is equipped with the unitary structure mentioned above [54]. Indeed, they constitute precisely the $S(U(2) \times U(2))$-finite vectors in this module.

Negative helicity fields: $k + 2 > 0$

The next instance of the transform is concerned with $\mathcal{O}(k)$ on $\mathbf{P}^I$, again, but with $k + 2 > 0$. We again recover an isomorphism between a cohomology group and a space of solutions of zero rest mass equations on $\mathbf{M}^I$ but with the opposite handedness, or helicity as the physicists call it. An interesting new feature is that the result of the transform is not directly a solution of the zero rest mass equations. Instead, we obtain a *potential* for a solution. This potential is a section of an auxiliary sheaf on Minkowski space. An actual zero rest mass field is obtained by applying an invariant differential operator occurring in a Bernstein–Gelfand–Gelfand resolution on $\mathbf{M}^I$. In fact, the potential comes naturally defined only up to an element of the kernel of this operator, usually called *gauge*. But since this ambiguity lies in the kernel of the invariant operator, the exactness of the resolution identifies the cohomology group and negative helicity zero rest mass fields.

Here are the details. Compute the relative BGG resolution on the fibres of η:

$$0 \rightarrow \eta^{-1}\mathcal{O}(k) \rightarrow \overset{k \quad 0 \quad 0}{\times\!\!-\!\!\times\!\!-\!\!\bullet} \rightarrow \overset{k+1 \;\; -2 \;\; 1}{\times\!\!-\!\!\times\!\!-\!\!\bullet} \rightarrow \overset{k+2 \;\; -3 \;\; 0}{\times\!\!-\!\!\times\!\!-\!\!\bullet} \rightarrow 0.$$

The case $k = -1$ is singular and is left to the reader. Suppose $k \geq 0$. Then the τ direct images of the sheaves of the resolution are non-zero only in degree zero:

$$\tau_*^0\left(\overset{k\quad 0\quad 0}{\times\!\!-\!\!\times\!\!-\!\!\bullet}\right) = \overset{k\quad 0\quad 0}{\bullet\!\!-\!\!\times\!\!-\!\!\bullet}$$

$$\tau_*^0\left(\overset{k+1\quad -2\quad 1}{\times\!\!-\!\!\times\!\!-\!\!\bullet}\right) = \overset{k+1\ -2\quad 1}{\bullet\!\!-\!\!\times\!\!-\!\!\bullet}$$

$$\tau_*^0\left(\overset{k+2\quad -3\quad 0}{\times\!\!-\!\!\times\!\!-\!\!\bullet}\right) = \overset{k+2\ -3\quad 0}{\bullet\!\!-\!\!\times\!\!-\!\!\bullet}.$$

The Leray spectral sequence is, again, a series of isomorphisms

$$H^q(\mathbf{F}^I, \Delta_\eta^p) \cong \Gamma(\mathbf{M}^I, \tau_*^q \Delta_\eta^p)$$

so that the hypercohomology sequence which computes $H^*(\mathbf{P}^I, \mathcal{O}(k))$ has an $E_1^{p,q}$ term with three terms in its lowest row and zero entries elsewhere:

$$E_1^{p,q}: \begin{array}{|ccc}
0 & 0 & 0 \\[4pt]
\overset{k\quad 0\quad 0}{\bullet\!\!-\!\!\times\!\!-\!\!\bullet} & \overset{k+1\ -2\ 1}{\bullet\!\!-\!\!\times\!\!-\!\!\bullet} & \overset{k+2\ -3\ 0}{\bullet\!\!-\!\!\times\!\!-\!\!\bullet}
\end{array}$$

The sequence collapses after one derivation:

$$E_2^{p,q} = E_\infty^{p,q}: \begin{array}{|ccc}
0 & 0 & 0 \\[6pt]
\overset{k\quad 0\quad 0}{\bullet\!\!-\!\!\bullet\!\!-\!\!\bullet} & \dfrac{\ker\left(\overset{k+1\ -2\ 1}{\bullet\!\!-\!\!\times\!\!-\!\!\bullet}\to\overset{k+2\ -3\ 0}{\bullet\!\!-\!\!\times\!\!-\!\!\bullet}\right)}{\mathrm{im}\left(\overset{k\quad 0\quad 0}{\bullet\!\!-\!\!\times\!\!-\!\!\bullet}\to\overset{k+1\ -2\ 1}{\bullet\!\!-\!\!\times\!\!-\!\!\bullet}\right)} & \text{coker}
\end{array}$$

Thus

$$H^1(\mathbf{P}^I, \mathcal{O}(k)) \cong \frac{\ker: \Gamma\left(\mathbf{M}^I, \overset{k+1\ -2\ 1}{\bullet\!\!-\!\!\times\!\!-\!\!\bullet}\right) \to \Gamma\left(\mathbf{M}^I, \overset{k+2\ -3\ 0}{\bullet\!\!-\!\!\times\!\!-\!\!\bullet}\right)}{\mathrm{im}: \Gamma\left(\mathbf{M}^I, \overset{k\quad 0\quad 0}{\bullet\!\!-\!\!\times\!\!-\!\!\bullet}\right) \to \Gamma\left(\mathbf{M}^I, \overset{k+2\ -3\ 0}{\bullet\!\!-\!\!\times\!\!-\!\!\bullet}\right)}$$

whilst (cf. remarks 9.2.1 and 9.2.2) the vanishing of $H^2(\mathbf{P}^I, \mathcal{O}(k))$ implies that the differential $\overset{k+1\ -2\ 1}{\bullet\!\!-\!\!\times\!\!-\!\!\bullet} \to \overset{k+2\ -3\ 0}{\bullet\!\!-\!\!\times\!\!-\!\!\bullet}$ is surjective on $\mathbf{M}^I$.

The differential operators appearing in the above passage come from the BGG resolution of $\overset{k\quad 0\quad 0}{\bullet\!\!-\!\!\bullet\!\!-\!\!\bullet}$ on $\mathbf{M}^I$. Using the technique of identifying their symbol they are readily computed as the (*higher valence*) *dual twistor operator* [128]

$$\pi_{\underbrace{B'\ldots C'}_{k\text{ terms}}} \to \nabla_{A(A'}\pi_{B'\ldots C')}$$

and an invariant operator

$$\Phi_{A\underbrace{B'C'\ldots D'}_{k+1\text{ terms}}} \mapsto \nabla_{A(A'}\Phi^A_{B'C'\ldots D')}.$$

Now the full BGG resolution of $\overset{k\quad 0\quad 0}{\bullet\!-\!\!\bullet\!-\!\!\bullet}$ is

$$\tag{15}$$

This is exact, so

$$\frac{\ker : \;\overset{k+1\;-2\;\;1}{\bullet\!-\!\times\!-\!\bullet} \to \overset{k+2\;-3\;\;0}{\bullet\!-\!\times\!-\!\bullet}}{\mathrm{im} : \;\overset{k\quad 0\quad 0}{\bullet\!-\!\times\!-\!\bullet} \to \overset{k+1\;-2\;\;1}{\bullet\!-\!\times\!-\!\bullet}} \cong \ker : \overset{0\;-k\text{-}3\;\;k+2}{\bullet\!-\!\times\!-\!\bullet} \to \overset{1\;-k\text{-}4\;\;k+1}{\bullet\!-\!\times\!-\!\bullet} \tag{16}$$

(This identification, using this resolution, first appeared in twistor theory in [30], before twistorians were aware of the full BGG resolution.)

The isomorphism is achieved by the mapping

$$\Phi_{AA'B'\ldots C'} \mapsto \underbrace{\nabla^{A'}_{(A}\nabla^{B'}_{B}\cdots\nabla^{C'}_{C}}_{k+1 \text{ terms}} \Phi_{D)A'B'\ldots C'} \tag{17}$$

applied to any representative of a class on the left-hand side of (16).

This (and a check on the singular case $k = -1$) establishes that, for $k > -2$,

$$H^1(\mathbf{P}^I, \mathcal{O}(k)) \cong \ker \;\; \Gamma(\mathbf{M}^I, \overset{0\;-k\text{-}3\;k+2}{\bullet\!-\!\times\!-\!\bullet}) \to \Gamma(\mathbf{M}^I, \overset{1\;-k\text{-}4\;k+1}{\bullet\!-\!\times\!-\!\bullet})$$

$$\cong \ker \;\; \nabla^{AA'} : \mathcal{O}_{\underbrace{(AB\ldots C)}_{k+2 \text{ terms}}}[-1] \to \mathcal{O}^{A'}_{\underbrace{(B\ldots C)}_{k+1 \text{ terms}}}[-3]$$

This is commonly called the space of negative helicity zero rest mass fields; the case $k = 0$, for instance, describes what physicists call *left-handed Maxwell fields*, on $\mathbf{M}^I$. The interesting new feature is that the transform no longer produces the physical fields directly. Instead (cf. the left-hand side of (16)), it gives *potentials* determined up to a freedom which physicists call a *gauge* freedom. These potentials must be differentiated (17) to obtain fields.

To illustrate this, concentrate on the case $k = 0$. The resolution (15) is the holomorphic de Rham resolution of $\mathbf{C}$, the operators in it being the usual exterior differentials on forms. The transform identifies

$$H^1(\mathbf{P}^I, \mathcal{O}) \cong \frac{\ker d^+ \; : \; \Gamma(\mathbf{M}^I, \Omega^1) \to \Gamma(\mathbf{M}^I, \Omega^2_+)}{\mathrm{im}\, d \; : \; \Gamma(\mathbf{M}^I, \mathcal{O}) \to \Gamma(\mathbf{M}^I, \Omega^1)}.$$

A one-form, Φ, representing a class of this group is a traditionally called a *potential* in electromagnetic theory. It is determined up to the addition of an exact one-form, which is called a *gauge* freedom. To compute an electromagnetic field, F, from a potential, we apply the exterior differential (which annihilates the gauge freedom, since the de Rham sequence is a complex). Φ satisfies $d^+\Phi = 0$ so that $F = d\Phi = d^-\Phi$ is an anti-self-dual two-form. By construction, F is exact, so $d*F = -dF = 0$; in other words, F is a left-handed electromagnetic field satisfying Maxwell's equations. Conversely, given any closed two-form $F \in \Gamma(\mathbf{M}^I, \Omega^2_-)$, the exactness of the de Rham resolution on $\mathbf{M}^I$ implies that $F = d^-\Phi$ for some one-form potential Φ satisfying $d^+\Phi = 0$ and determined up to the gauge freedom of adding an exact one-form. This establishes that

$$H^1(\mathbf{PT}^I, \mathcal{O}) \cong \ker d \ : \ \Gamma(\mathbf{M}^I, \Omega^2_-) \to \Gamma(\mathbf{M}^I, \Omega^3).$$

It is this case that admits a non-linear version known as the *Ward correspondence* which gives a twistorial account of anti-self-dual Yang–Mills fields (see [153] and section 9.9).

Remark (9.2.6).   The construction has been carried out explicitly for cohomology on $\mathbf{P}^I$, but extends, as before, to $\overline{\mathbf{PT}^+}$ and $\overline{\mathbf{P}^-}$, etc., to give, respectively, positive and negative frequency, negative helicity, massless fields.

Remark (9.2.7).   For these left-handed fields, the analytic properties of the chosen $U \subset \mathbf{M}$ are crucial (in addition to the topological properties (cf. remark 9.2.4)). Not only is the Leray spectral sequence affected, but also the identification on U of potentials modulo gauge as fields. This boils down to being able to compute the topological cohomology of U by means of the BGG resolution, a fact which relies on Cartan's theorem B. For example, it may be impossible to find a holomorphic potential for an anti-self-dual holomorphic Maxwell field even though there is no topological obstruction to so doing. Indeed, in [29] Buchdahl exhibits a *contractible* open subset $U \subset \mathbf{M}$ for which there is an exact sequence

$$\left[\frac{\ker d^+ : \Gamma(U, \Omega^1) \to \Gamma(U, \Omega^2_+)}{\operatorname{im} d : \Gamma(U, \mathcal{O}) \to \Gamma(U, \Omega^1)} \right] \hookrightarrow \left[\ker d : \Gamma(U, \Omega^2_-) \to \Gamma(U, \Omega^3) \right] \to \mathbf{C} \to 0.$$

A complete proof involves some detailed hard analysis but the rough idea is as follows. The first two terms are unchanged if U is replaced by $\hat{U}$, its envelope of holomorphy. Specifically, we choose

$$\hat{U} = \left(\begin{array}{c} \text{A Stein ``fattening'' in } \mathbf{C}^3 \\ \text{of the sphere } S^2 \subset \mathbf{R}^3 \subset \mathbf{C}^3 \end{array} \right) \times \{|z_4| < 1\}$$

and

$$U = \{z \in \hat{U} \text{ s.t. } (x_1, x_2) \neq (0,0) \text{ and } x_3 > 0\}.$$

Since holomorphic functions extend uniquely across a non-complex codimension two real analytic subvariety it is clear that $\hat{U}$ is the envelope of holomorphy of U. By construction, $\hat{U}$ has the homotopy type of the sphere S^2 whereas U has the homotopy type of a punctured sphere, i.e., is contractible. The $\mathbf{C}$ which occurs in the exact sequence above now comes from $H^2(\hat{U}, \mathbf{C})$ and the remaining analysis of [29] is to show that, by a suitable choice of $\hat{U}$, the natural mapping

$$\left[\ker d : \Gamma(U, \Omega_-^2) \to \Gamma(U, \Omega^3)\right] \to \mathbf{C}$$

is surjective as claimed. A further discussion is in [29,44].

The scalar wave equation: $k = -2$

The third historical case of the Penrose transform interprets the cohomology of $\mathcal{O}(-2)$ on $\mathbf{P}^I$ as solutions of the massless free field wave equation on $\mathbf{M}^I$. From the point of view of representation theory, it is perhaps the most interesting case. Because the weight $\overset{-2}{\bullet}\,\overset{0}{\rule{0pt}{0pt}}\!\!\!-\!\!\!\overset{0}{\bullet}$ is "very singular," the hypercohomology spectral sequence takes longer to converge. The most immediate consequence of this is that first-order differential operators in the fibrewise resolution induce a second-order differential operator between sections of homogeneous sheaves on $\mathbf{M}^I$ where a first-order operator has hitherto arisen. Higher order operators have already appeared above, but these came always from higher order operators in the Bernstein–Gelfand–Gelfand resolution.

Resolving, as usual, we obtain

$$0 \to \eta^{-1}\,\overset{-2\ \ \ 0\ \ \ 0}{\underset{\times\!-\!\bullet\!-\!\bullet}{}} \to \overset{-2\ \ \ 0\ \ \ 0}{\underset{\times\!-\!\times\!-\!\bullet}{}} \to \overset{-1\ -2\ \ \ 1}{\underset{\times\!-\!\times\!-\!\bullet}{}} \to \overset{0\ -3\ \ \ 0}{\underset{\times\!-\!\times\!-\!\bullet}{}} \to 0.$$

Direct images are computed as

$$\tau_*^1\,\overset{-2\ \ \ 0\ \ \ 0}{\underset{\times\!-\!\times\!-\!\bullet}{}} = \overset{0\ -1\ \ \ 0}{\underset{\bullet\!-\!\times\!-\!\bullet}{}}$$

$$\tau_*^0\,\overset{0\ -3\ \ \ 0}{\underset{\times\!-\!\times\!-\!\bullet}{}} = \overset{0\ -3\ \ \ 0}{\underset{\bullet\!-\!\times\!-\!\bullet}{}}$$

and all others vanish. In particular, this means that there is a column of zeros in the $E_1^{p,q}$ term of the hypercohomology spectral sequence, where previously at least one non-zero entry occurred:

$$E_1^{p,q} : \left[\begin{array}{ccccc} \overset{0 \;\; -1 \;\; 0}{\bullet\!\!-\!\!\overset{\times}{\longrightarrow}\!\!-\!\!\bullet} & 0 & & 0 & \\ & 0 & 0 & \overset{0 \;\; -3 \;\; 0}{\bullet\!\!-\!\!\overset{\times}{\longrightarrow}\!\!-\!\!\bullet} \end{array} \right.$$

Consequently, $E_1^{p,q} = E_2^{p,q}$ and the sequence converges on the second derivation to

$$E_3^{p,q} = E_\infty^{p,q} : \left| \begin{array}{ccc} \mathrm{ker} : \overset{0 \;\; -1 \;\; 0}{\bullet\!\!-\!\!\overset{\times}{\longrightarrow}\!\!-\!\!\bullet} \to \overset{0 \;\; -3 \;\; 0}{\bullet\!\!-\!\!\overset{\times}{\longrightarrow}\!\!-\!\!\bullet} & 0 & 0 \\ 0 & 0 \;\; \mathrm{coker} : \overset{0 \;\; -1 \;\; 0}{\bullet\!\!-\!\!\overset{\times}{\longrightarrow}\!\!-\!\!\bullet} \to \overset{0 \;\; -3 \;\; 0}{\bullet\!\!-\!\!\overset{\times}{\longrightarrow}\!\!-\!\!\bullet} \end{array} \right.$$

This time, therefore, the differential operator induced down on $\mathbf{M}^I$ arises as a d_2 differential at the second level of the spectral sequence. To compute the operator (up to scale), we recall again that

$$\odot^2 \Omega_{\mathbf{M}^I}^1 = \overset{2 \;\; -4 \;\; 2}{\bullet\!\!-\!\!\overset{\times}{\longrightarrow}\!\!-\!\!\bullet} \oplus \overset{0 \;\; -2 \;\; 0}{\bullet\!\!-\!\!\overset{\times}{\longrightarrow}\!\!-\!\!\bullet}$$

so that the symbol of the operator is the projection

$$\overset{0 \;\; -1 \;\; 0}{\bullet\!\!-\!\!\overset{\times}{\longrightarrow}\!\!-\!\!\bullet} \otimes (\odot^2 \Omega_{\mathbf{M}^I}^1) = \overset{2 \;\; -5 \;\; 2}{\bullet\!\!-\!\!\overset{\times}{\longrightarrow}\!\!-\!\!\bullet} \oplus \overset{0 \;\; -3 \;\; 0}{\bullet\!\!-\!\!\overset{\times}{\longrightarrow}\!\!-\!\!\bullet}$$
$$\to \overset{0 \;\; -3 \;\; 0}{\bullet\!\!-\!\!\overset{\times}{\longrightarrow}\!\!-\!\!\bullet}$$

(we readily check that there is no other possibility). This identifies the operator as the *wave operator*, $\square = \nabla^a \nabla_a$, acting on functions with conformal weight $[-1]$ and so,

$$H^1(\mathbf{P}^I, \mathcal{O}(-2)) \cong \mathrm{ker} \;\square : \Gamma(\mathbf{M}^I, \overset{0 \;\; -1 \;\; 0}{\bullet\!\!-\!\!\overset{\times}{\longrightarrow}\!\!-\!\!\bullet}) \to \Gamma(\mathbf{M}^I, \overset{0 \;\; -3 \;\; 0}{\bullet\!\!-\!\!\overset{\times}{\longrightarrow}\!\!-\!\!\bullet})$$
$$\cong \mathrm{ker} \;\square : \Gamma(\mathbf{M}^I, \mathcal{O}[-1]) \to \Gamma(\mathbf{M}^I, \mathcal{O}[-3]) .$$

Remark (9.2.8).   In the introduction to this chapter we remarked that the translation invariant differential operators produced by the Penrose transform are dual to homomorphisms of (generalized) Verma modules. The wave operator is our first example of an operator constructed by the transform which is related to a *non-standard* homomorphism of Verma modules. Geometrically, this is a consequence of the fact that $\square$ is not a direct image of an invariant differential operator on the intermediate space $\mathbf{F}^I$.

Remark (9.2.9).   It is quite easy to compute the d_2 differential in the hypercohomology spectral sequence.

Recall, firstly, the construction of the hypercohomology spectral sequence for a resolution

$$0 \to \mathcal{F} \to \Delta_\eta^\bullet \qquad\qquad (18)$$

of coherent sheaves where the sheaves Δ_η^i may have non-vanishing higher cohomology. We shall, for definiteness, work with Čech cohomology, but of course we could equally work abstractly with injective resolutions or with Dolbeault cohomology. So let $\mathcal{U}$ be a good Stein (or affine) cover of the ambient space, X. This means that for all coherent sheaves $\mathcal{G}$, $H^i(U, \mathcal{G}) = 0$ for $i > 0$ where U is a (finite) intersection of members of $\mathcal{U}$. Let $C^\bullet(\mathcal{U}, \mathcal{G})$ be the usual Čech cochain complex of $\mathcal{G}$ [80]. The differentials d of the resolution (18) and δ of the complexes $C^\bullet(\mathcal{U}, \Delta_\eta^i)$ make

$$E_0^{p,q} = C^p(\mathcal{U}, \Delta_\eta^q)$$

a double complex [25]. Deriving with respect to d, first, gives

$$E_1^{p,q} = \begin{cases} C^p(\mathcal{U}, \mathcal{F}) & q = 0 \\ 0 & \text{otherwise} \end{cases}$$

and then, deriving with respect to δ,

$$E_2^{p,q} = \begin{cases} H^p(X, \mathcal{F}) & q = 0 \\ 0 & \text{otherwise} \end{cases}$$

since $\mathcal{U}$ is a good cover. This spectral sequence has converged to the total cohomology of $E_0^{p,q}$, which is therefore just the cohomology of $\mathcal{F}$ on X. Alternatively, deriving first with respect to δ gives

$$E_1^{p,q} = H^q(X, \Delta_\eta^p)$$

which is just the first term of the hypercohomology spectral sequence. By the general theory of spectral sequences, this must converge, also, to the total cohomology of $E_0^{p,q}$, i.e., to the cohomology of $\mathcal{F}$.

The differentials in $H^q(X, \Delta_\eta^p)$ are induced from the d of the resolution (18). We are interested in deriving once again and computing the differentials d_2 in the $E_2^{p,q}$ term. A detailed description of what follows is given in [25, on p. 161ff]. This differential maps $d_2 : E_2^{p,q} \to E_2^{p+2,q-1}$. A little thought shows that a class $[\alpha] \in E_2^{p,q}$ may be represented by an element $\alpha \in C^q(\mathcal{U}, \Delta_\eta^p)$ satisfying

   (i) $\delta\alpha = 0$

   (ii) $d\alpha = \delta\beta$

where $\beta \in C^{q-1}(\mathcal{U}, \Delta_\eta^{p+1})$. The first condition expresses the fact that α must represent a cocycle in $E_1^{p,q}$ and the second the fact that this cocycle lies in the kernel of the induced differential $d : E_1^{p,q} \to E_1^{p+1,q}$, so that $[\alpha]$ is a class in $E_2^{p,q}$. Then,

$$d_2[\alpha] = [d\beta]$$

(standard "diagram chasing" ensures that d_2 is well defined and independent of any choices of class representatives).

Now we have a sufficiently explicit description of the workings of the hypercohomology spectral sequence to calculate directly the wave operator as a d_2 map. We wish to compute cohomology on the intermediate space $\mathbf{F}^I \subset$ ×—×—• which, we observed, is the projectivization of the spinor bundle $\overset{1}{\bullet}\!\!-\!\!\overset{-1}{\times}\!\!-\!\!\overset{0}{\bullet}$ on $\mathbf{M}^I$. As a complex manifold, the intermediate space is a trivial product $\mathbf{CP}^1 \times \mathbf{M}^I$, for $\overset{1}{\bullet}\!\!-\!\!\overset{-1}{\times}\!\!-\!\!\overset{0}{\bullet}$ admits a basis of non-vanishing sections over $\mathbf{M}^I$. To see this, recall that $\mathbf{P}^I = \mathbf{CP}^3 \setminus L$ where L is the line in Twistor space corresponding to the light cone at infinity in $\mathbf{CS}^4$. Fix $A^\alpha, B^\alpha \in \mathbf{C}^4$ whose span gives L. A choice of flat metric on $\mathbf{M}^I$ gives a splitting of the sequence

$$0 \to \overset{1}{\bullet}\!\!-\!\!\overset{-1}{\times}\!\!-\!\!\overset{0}{\bullet} \to \overset{0}{\bullet}\!\!-\!\!\overset{0}{\bullet}\!\!-\!\!\overset{1}{\bullet} \to \overset{0}{\bullet}\!\!-\!\!\overset{0}{\times}\!\!-\!\!\overset{1}{\bullet} \to 0$$

so we let $o_{A'}, \iota_{A'} \in \Gamma(\mathbf{M}^I, \overset{1}{\bullet}\!\!-\!\!\overset{-1}{\times}\!\!-\!\!\overset{0}{\bullet})$ correspond to A^α, B^α under this splitting. They constitute the required basis, and are covariant constant with respect to the Levi-Civita connection of the metric. Let $[\pi_{A'}]$ be homogeneous co-ordinates on the factor $\mathbf{CP}^1$ (which may be thought of as covariantly constant spinor fields on $\mathbf{M}^I$.)

A choice of metric fixes also two covariantly constant antisymmetric spinors $\epsilon^{A'B'}$ and ϵ^{AB} and a means of raising indices:

$$o^{A'} = \epsilon^{A'B'} o_{B'}; \iota^{A'} = \epsilon^{A'B'} \iota_{B'}; \nabla^A_{A'} = \epsilon^{AB} \nabla_{BA'}, \text{ etc.}$$

Normalize so that $o_{A'}\iota^{A'} = 1$. This gives

$$\pi_{A'} = (\pi_{C'}\iota^{C'})o_{A'} - (\pi_{C'}o^{C'})\iota_{A'}.$$

Then there is a good Stein cover of $\mathbf{F}^I$, $\mathcal{U}$ by two sets:

$$\{[\pi_{A'}]; \pi_{A'}\iota^{A'} \neq 0\} \times \mathbf{M}^I$$

and

$$\{[\pi_{A'}]; \pi_{A'}o^{A'} \neq 0\} \times \mathbf{M}^I.$$

The resolution of $\eta^{-1} \overset{-2}{\times}\!\!-\!\!\overset{0}{\bullet}\!\!-\!\!\overset{0}{\bullet}$, with its differentials, is

$$0 \to \eta^{-1}\overset{-2}{\times}\!\!-\!\!\overset{0}{\bullet}\!\!-\!\!\overset{0}{\bullet} \to \overset{-2}{\times}\!\!-\!\!\overset{0}{\times}\!\!-\!\!\overset{0}{\bullet} \xrightarrow{\pi^{A'}\nabla_{AA'}} \overset{-1}{\times}\!\!-\!\!\overset{-2}{\times}\!\!-\!\!\overset{1}{\bullet} \xrightarrow{\pi^{B'}\nabla^A_{B'}} \overset{0}{\times}\!\!-\!\!\overset{-3}{\times}\!\!-\!\!\overset{0}{\bullet} \to 0.$$

Remark (9.2.10).   The composition $\nabla_{A(A'}\nabla^A_{B')}$ is the square of the exterior differential on functions (given the choice of metric), hence zero.

Any element of $H^1(\mathbf{F}^I, \overset{\text{-2}}{\times}\!\!-\!\!\overset{0}{\times}\!\!-\!\!\overset{0}{\bullet})$ can be represented by a two-cocycle of the form

$$\alpha = \frac{g}{(\pi_{C'}o^{C'})(\pi_{D'}\iota^{D'})}$$

where $g \in \Gamma(\mathbf{M}^I, \overset{0}{\bullet}\!\!-\!\!\overset{\text{-1}}{\times}\!\!-\!\!\overset{0}{\bullet})$ (by the Leray sequence and the Bott–Borel–Weil theorem). This satisfies condition (i), above, trivially, since our cover has only two sets. Applying the first differential of the resolution, obtain

$$d\alpha = \frac{\pi^{A'}\nabla_{AA'}g}{(\pi_{C'}o^{C'})(\pi_{D'}\iota^{D'})}$$

$$= \frac{-\iota^{A'}\nabla_{AA'}g}{\pi_{C'}\iota^{C'}} + \frac{o^{A'}\nabla_{AA'}g}{\pi_{C'}o^{C'}}$$

This clearly has the desired form, $\delta\beta$, where

$$\beta = \left(\frac{\iota^{A'}\nabla_{AA'}g}{\pi_{C'}\iota^{C'}}, \frac{o^{A'}\nabla_{AA'}g}{\pi_{C'}o^{C'}}\right)$$

is a one-cocycle relative to $\mathcal{U}$. Then

$$d_2[\alpha] = [d\beta]$$

$$= \frac{\iota^{A'}\pi^{B'}\nabla^A_{A'}\nabla_{AB'}g}{\pi_{C'}\iota^{C'}}$$

$$= \iota^{A'}o^{B'}\nabla^A_{A'}\nabla_{AB'}g$$

$$= \nabla^{AA'}\nabla_{AA'}g$$

$$= \Box g$$

(a reader unfamiliar with the abstract index notation and spinor algebra used here should consult [127]). An alternative argument for identifying $\Box$, midway in abstraction between the two given above, is provided in [44].

9.3 The Penrose transform of forms on twistor space

It is now time to apply the Penrose transform to some homogeneous vector bundles, rather than simply to line bundles. We shall see that the Penrose transform for vector bundles often yields rather more complicated results than those obtained so far for line bundles. Up to now, cohomology groups have been identified with solutions of differential equations (perhaps after slight reinterpretation). They have, in fact, been irreducible modules for $\mathbf{g} = \mathbf{sl}(4, \mathbf{C})$ as we shall see in section 11.4. This is not generically true; to see examples of the more general behaviour of the transform we shall study its application to differential forms on $\mathbf{P}$. The result is typical of what is obtained when the transform is applied to a sheaf $\mathcal{O}_{\mathbf{P}}(\lambda)$ where λ is non-singular but neither dominant nor antidominant for $\mathbf{g}$. See [51] for an earlier version of these calculations.

Consider $\Omega_{\mathbf{P}}^1$, first. The Bernstein–Gelfand–Gelfand resolution is

$$0 \to \eta^{-1} \overset{-2 \quad 1 \quad 0}{\underset{\times \!\!-\!\! \bullet \!\!-\!\! \bullet}{}} \to \overset{-2 \quad 1 \quad 0}{\underset{\times \!\!-\!\! \times \!\!-\!\! \bullet}{}} \to \overset{0 \quad -3 \quad 2}{\underset{\times \!\!-\!\! \times \!\!-\!\! \bullet}{}} \to \overset{1 \quad -4 \quad 1}{\underset{\times \!\!-\!\! \times \!\!-\!\! \bullet}{}} \to 0$$

and the first term of the hypercohomology spectral sequence is

$$E_1^{p,q} \left| \begin{array}{ccc} \mathcal{O} & 0 & 0 \\ 0 & \Omega_-^2 & \Omega^3 \end{array} \right.$$

(where, as before, the individual entries in this term stand for the space of sections of the indicated bundles over the affine set $\mathbf{M}^I$). Deriving once, using (remark 9.2.2) the exactness of

$$\Gamma(\mathbf{M}^I, \Omega_-^2) \to \Gamma(\mathbf{M}^I, \Omega^3) \to \Gamma(\mathbf{M}^I, \Omega^4) \to 0$$

on $\mathbf{M}^I$, we obtain

$$E_2^{p,q} \left| \begin{array}{ccc} \mathcal{O} & 0 & 0 \\ 0 & \ker d : \Omega_-^2 \to \Omega^3 & \Omega^4 \end{array} \right.$$

Observing, yet again, that $H^2(\mathbf{P}^I, \Omega^1) = 0$, we deduce that the operator

$$d_2 : \Gamma(\mathbf{M}^I, \mathcal{O}) \to \Gamma(\mathbf{M}^I, \Omega^4)$$

is surjective. Checking possible symbols, we quickly find that d_2 is proportional to the square of the Laplacian, $\square^2 = (\nabla . \nabla)^2$, where ∇ is the Levi–Civita connection of any flat metric in the conformal class of metrics on $\mathbf{M}^I$. In particular, this operator is invariant under conformal motions on $\mathbf{M}^I$. It follows that

$$0 \to \{\ker d : \Gamma(\mathbf{M}^I, \Omega_-^2) \to \Gamma(\mathbf{M}^I, \Omega^3)\} \to H^1(\mathbf{P}^I, \Omega_{\mathbf{P}}^1)$$
$$\to \Gamma(\mathbf{M}^I, \mathcal{O}) \overset{\square^2}{\to} \Gamma(\mathbf{M^I}, \Omega^4) \to 0$$

is exact. Notice that the first term of this sequence is exactly the space of anti-self-dual Maxwell fields on $\mathbf{M}^I$ which, as on page 103, may be represented as $H^1(\mathbf{P}^I, \mathcal{O})$ on twistor space. The mapping to $H^1(\mathbf{P}^I, \Omega^1)$ is induced by the exterior differential $d : \mathcal{O} \to \Omega^1$. $\square^2$ is another example of a non-standard differential operator, coming from a non-standard homomorphism of Verma modules.

Next, consider $\Omega_{\mathbf{P}}^2 = \overset{-3\ \ \ 0\ \ \ 1}{\times\!\!-\!\!\bullet\!\!-\!\!\bullet}$. This has a BGG resolution

$$0 \to \eta^{-1} \overset{-3\ \ 0\ \ 1}{\times\!-\!\bullet\!-\!\bullet} \to \overset{-3\ \ 0\ \ 1}{\times\!-\!\times\!-\!\bullet} \to \overset{-2\ \ -2\ \ 2}{\times\!-\!\times\!-\!\bullet} \to \overset{0\ \ -4\ \ 0}{\times\!-\!\times\!-\!\bullet} \to 0.$$

The first term in the hypercohomology spectral sequence is

$$E_1^{p,q} \quad \begin{array}{|ccc} \Omega^1 & \Omega_-^2 & 0 \\ 0 & 0 & \Omega^4 \end{array}$$

Deriving, we obtain

$$E_2^{p,q} \quad \begin{array}{|ccc} \{\ker d_- : \Omega^1 \to \Omega_-^2\} & 0 & 0 \\ 0 & & 0 \quad \Omega^4 \end{array}$$

The vanishing of second cohomology again implies that

$$\{\ker \Gamma(\mathbf{M}^I, \Omega^1) \to \Gamma(\mathbf{M}^I, \Omega_-^2)\} \to \Gamma(\mathbf{M}^I, \Omega^4)$$

is surjective. Hence

$$0 \to H^1(\mathbf{P}^I, \Omega_{\mathbf{P}}^2) \to \{\ker \Gamma(\mathbf{M}^I, \Omega^1) \to \Gamma(\mathbf{M}^I, \Omega_-^2)\} \to \Gamma(\mathbf{M}^I, \Omega^4) \to 0$$

is exact. This may be rephrased by using the following sequence, deduced from the exactness of the BGG resolution of $\mathbf{C}$ on $\bullet\!\!-\!\!\times\!\!-\!\!\bullet$

$$0 \to \mathbf{C} \to \mathcal{O} \overset{d}{\to} \{\ker \Gamma(\mathbf{M}^I, \Omega^1) \to \Gamma(\mathbf{M}^I, \Omega_-^2)\}$$
$$\to \{\ker \Gamma(\mathbf{M}^I, \Omega_+^2) \to \Gamma(\mathbf{M}^I, \Omega^3)\} \to 0$$

and the fact that $\square^2$ factors through $\mathcal{O} \overset{d}{\to} \Omega^1$ to deduce an exact sequence

$$0 \to \frac{\square^2 : \Gamma(\mathbf{M}^I, \mathcal{O}) \to \Gamma(\mathbf{M}^I, \Omega^4)}{\mathbf{C}} \to H^1(\mathbf{P}^I, \Omega_{\mathbf{P}}^2)$$
$$\to \{\ker (\Gamma(\mathbf{M}^I, \Omega_+^2) \to \Gamma(\mathbf{M}^I, \Omega^3)\} \to 0.$$

In particular, since the maps in this sequence intertwine the action of $\mathrm{sl}(4, \mathbf{C})$, $H^1(\mathbf{P}^I, \Omega^2)$ is not irreducible.

Notice also the space of self-dual Maxwell fields in this sequence and recall, as on page 99, that this space is the Penrose transform of $H^1(\mathbf{P}^I, \Omega^3)$. Thus we see that the whole sequence

$$H^1(\mathbf{P}^I, \mathcal{O}) \to H^1(\mathbf{P}^I, \Omega^1) \to H^1(\mathbf{P}^I, \Omega^2) \to H^1(\mathbf{P}^I, \Omega^3)$$

has been Penrose transformed to $\mathbf{M}^I$.

9.4   Other bundles on twistor space

All other non-singular homogeneous bundles on $\mathbf{P}$ will transform according to a pattern similar to one of the form bundles (including $\overset{0}{\times}\!\!-\!\!\overset{0}{\bullet}\!\!-\!\!\overset{0}{\bullet}$ and $\overset{-4}{\times}\!\!-\!\!\overset{0}{\bullet}\!\!-\!\!\overset{0}{\bullet} = \Omega_{\mathbf{P}}^3$). Nonetheless, it is worth while to compute some further examples which have particular significance in twistor theory.

Deformations and nonlinear gravitons

One of the most significant results of twistor theory, obtained by Penrose in [123], is the characterization of all anti-self-dual conformal manifolds in a neighbourhood of Minkowski space using deformations of $\mathbf{P}^I$, or, at least, of some neighbourhood of a line in $\mathbf{P}^I$. The first step in this construction is to compute the (Zariski) tangent space to the moduli space of such conformal manifolds at Minkowski space. Penrose showed that this is equivalent to computing the (Zariski) tangent space to the space of deformations of $\mathbf{P}^I$ at $\mathbf{P}^I$.

This can be computed, following the work of Kodaira and Spencer [101], as the cohomology group $H^1(\mathbf{P}^I, \Theta)$ where Θ is the holomorphic tangent bundle of $\mathbf{P}^I$. (For a brief sketch of why this should be so, see [12,33].) Furthermore, there may exist obstructions to a particular direction in the Zariski tangent space corresponding to an actual finite curve in the moduli space. Indeed, an infinite sequence of such obstructions exists, each lying in $H^2(\mathbf{P}^I, \Theta)$, according to the general theory. Fortunately for twistor theory, this group is zero, so that by (a small modification of) the work of Kodaira and Spencer, there is no obstruction to the exponentiation of these infinitesimal deformations.

Now $\Theta = \overset{1}{\times}\!\!-\!\!\overset{0}{\bullet}\!\!-\!\!\overset{1}{\bullet}$; the appropriate BGG resolution is

$$0 \to \overset{1}{\times}\!\!-\!\!\overset{0}{\bullet}\!\!-\!\!\overset{1}{\bullet} \to \overset{1}{\times}\!\!-\!\!\overset{0}{\times}\!\!-\!\!\overset{1}{\bullet} \to \overset{2}{\times}\!\!-\!\!\overset{-2}{\times}\!\!-\!\!\overset{2}{\bullet} \to \overset{4}{\times}\!\!-\!\!\overset{-4}{\times}\!\!-\!\!\overset{0}{\bullet} \to 0.$$

The first term of the hypercohomology spectral sequence is

$$
\begin{array}{c|ccc}
E_1^{p,q} & 0 & 0 & 0 \\
\hline
& \overset{1}{\bullet}\!\!-\!\!\overset{0}{\times}\!\!-\!\!\overset{1}{\bullet} & \overset{2}{\bullet}\!\!-\!\!\overset{-2}{\times}\!\!-\!\!\overset{2}{\bullet} & \overset{4}{\bullet}\!\!-\!\!\overset{-4}{\times}\!\!-\!\!\overset{0}{\bullet}
\end{array}
$$

This yields, in "potential modulo gauge" form,

$$H^1(\mathbf{P}^I, \Theta) \cong \frac{\ker\ \overset{2}{\bullet}\!\!-\!\!\overset{-2}{\times}\!\!-\!\!\overset{2}{\bullet} \to \overset{4}{\bullet}\!\!-\!\!\overset{-4}{\times}\!\!-\!\!\overset{0}{\bullet}}{\mathrm{im}\ \overset{1}{\bullet}\!\!-\!\!\overset{0}{\times}\!\!-\!\!\overset{1}{\bullet} \to \overset{2}{\bullet}\!\!-\!\!\overset{-2}{\times}\!\!-\!\!\overset{2}{\bullet}}$$

$$\cong \ker\ \overset{0}{\bullet}\!\!-\!\!\overset{-4}{\times}\!\!-\!\!\overset{4}{\bullet} \to \overset{2}{\bullet}\!\!-\!\!\overset{-6}{\times}\!\!-\!\!\overset{2}{\bullet}$$

which identifies the infinitesimal deformations of $\mathbf{P}^I$ with linearized anti-self-dual deformations of Minkowski space.

The Einstein bundle

A closely related bundle is $\mathcal{E} = \overset{0}{\times}\!\!-\!\!\overset{1}{\bullet}\!\!-\!\!\overset{0}{\bullet}$. The space of sections of $\mathcal{E}$ is identified by the Penrose transform as

$$\Gamma(\mathbf{P}, \mathcal{E}) \cong \ker \Gamma(\mathbf{M}, \overset{0}{\bullet}\!\!-\!\!\overset{1}{\times}\!\!-\!\!\overset{0}{\bullet}) \overset{D}{\to} \Gamma(\mathbf{M}, \overset{2}{\bullet}\!\!-\!\!\overset{-3}{\times}\!\!-\!\!\overset{2}{\bullet}). \tag{19}$$

Up to now, we have expressed the differential operators in the Penrose transform using the Levi–Civita connection of a flat metric in the conformal class. They can be expressed in terms of the Levi–Civita connection of any metric in the conformal class. Letting

$$2\Phi_{ab} = -(R_{ab} - \tfrac{1}{4}g_{ab}R)$$

for R_{ab}, R, the Ricci and scalar curvatures of the metric g_{ab}, we obtain

$$\mathbf{D} : f \to \{\nabla_{(A'}^{(A}\nabla_{B')}^{B)} + \Phi_{A'B'}^{AB}\}f. \tag{20}$$

However, $\mathcal{E} = \Omega^1 \otimes \mathcal{O}(2)$ gives an intrinsic definition of $\mathcal{E}$ even on deformations $\mathcal{P}$ of $\mathbf{P}^I$. There is a curved space version of the Penrose transform in which (19), with $\mathbf{D}$ as in (20), remains valid [8,11,58,70]. Then, if $\Gamma(\mathcal{P}, \mathcal{E}) \neq 0$, there is a non-zero section $f \in \overset{0}{\bullet}\!\!-\!\!\overset{1}{\times}\!\!-\!\!\overset{0}{\bullet}$, the conformal weight one line bundle on the anti-self-dual conformal manifold $\mathcal{M}$ corresponding to $\mathcal{P}$. Rescale the metric by $g_{ab} \to \hat{g}_{ab} = f^{-2}g_{ab}$ replacing $\Phi_{ab} \to \hat{\Phi}_{ab}$, etc., and $f \to \hat{f} = f^{-1}f = 1$. Then

$$0 = \{\hat{\nabla}_{(A'}^{(A}\hat{\nabla}_{B')}^{B)} + \hat{\Phi}_{A'B'}^{AB}\}\hat{f} = \hat{\Phi}_{A'B'}^{AB}$$

and $\hat{g}_{ab}$ is an *Einstein* metric on $\mathcal{M}$. The non-vanishing of $\Gamma(\mathcal{P}, \mathcal{E})$ is consequently of differential-geometric and physical significance [57,106,154].

9.5   The Penrose transform for ambitwistor space

The Penrose transform interpretation of zero rest mass fields has the rather interesting feature that it treats fields of positive or negative helicity in an essentially different way. So, for example, left- and right-handed Maxwell fields are obtained from distinct cohomology groups. This may be an advantage in trying to build a physical theory using the transform, for nature iself appears to be biased between left- and right-handedness. From the representation theory point of view the situation is clear: different helicities lie in distinct cohomology groups on twistor space because (see 11.4) they correspond to distinct irreducible representations. On the other hand, mixed helicity fields often occur in nature so we should find a way of using the Penrose transform to describe them. The answer is to study an *obstruction* problem on ambitwistor space. Exactly why this works from the representation point of view will be explained below.

Consider next the double fibration

The fibres of η are one dimensional so that points of ✕—●—✕ correspond to lines in ●—✕—● which are *null* in the conformal structure of ●—✕—●, and all null lines arise this way (see example 2.3.2). The manifold ✕—●—✕ is often called ambitwistor space and denoted $\mathbf{A}$. The open subset corresponding to $\mathbf{M}^I$ is denoted $\mathbf{A}^I$.

Because the η-fibres are one dimensional, BGG resolutions in the Penrose transform are short exact sequences and the long exact sequences in cohomology which they induce amount to the hypercohomology spectral sequence. Consider, for example, $H^i(\mathbf{A}^I, \mathcal{O})$. The BGG resolution is

$$0 \to \underset{\times\!-\!\bullet\!-\!\times}{\overset{0\quad 0\quad 0}{}} \to \underset{\times\!-\!\times\!-\!\times}{\overset{0\quad 0\quad 0}{}} \to \underset{\times\!-\!\times\!-\!\times}{\overset{1\quad\text{-}2\quad 1}{}} \to 0.$$

Taking the long exact sequence in cohomology on $\tau^{-1}\mathbf{M}$ (and using the degenerate Leray spectral sequence as usual), obtain

$$\Gamma(\mathbf{A}^I, \mathcal{O}) \cong \mathbf{C} \quad \text{and} \quad H^1(\mathbf{A}^I, \mathcal{O}) \cong \text{coker } \Gamma(\mathbf{M}^I, \mathcal{O}) \overset{d}{\to} \Gamma(\mathbf{M}^I, \Omega^1).$$
$$\cong \text{ker } \Gamma(\mathbf{M}^I, \Omega^2) \overset{d}{\to} \Gamma(\mathbf{M}^I, \Omega^3)$$

This is not quite as satisfactory as we might have hoped, from the physical point of view. Contained in $\Gamma(\mathbf{A}^I, \mathcal{O})$ are the Maxwell fields

$$\{F \in \Gamma(\mathbf{M}^I, \Omega^2) | d * F = dF = 0\}$$

where $*$ is the Hodge star operator of a flat metric on $\mathbf{M}^I$. These represent free electromagnetic fields on $\mathbf{M}^I$. These may be singled out using an *extension problem* for the embedding $\mathbf{A}^I \hookrightarrow \mathbf{P}^I \times \mathbf{P}^{*I}$ as first observed by Green, Isenberg and Yasskin [72] and Witten [161] with cohomological interpretation by Henkin and Manin [84,113], Buchdahl [32] and Pool [129]. Here, $\mathbf{P}^{*I} \subset$ •———•——× corresponds to $\mathbf{M}^I$ and the embedding comes from the projections

$$×——•——• \leftarrow ×——•——× \rightarrow •——•——×.$$

Let $\mathcal{I}$ denote the *ideal sheaf* of this embedding. Define $\mathcal{O}_{(k)} = \mathcal{O}_{\mathbf{P} \times \mathbf{P}^*}/\mathcal{I}^{k+1}$. Sections of this sheaf correspond to functions determined to order k off $\mathbf{A}^I$. (The ringed space $(\mathbf{A}^I, \mathcal{O}_{(k)})$ is called the kth *formal neighbourhood* of $\mathbf{A}^I$). Then $\mathcal{I}^k/\mathcal{I}^{k+1} \cong$ ×$\overset{-k}{\text{———}}$•$\overset{0}{\text{———}}$×$\overset{-k}{}$ is the kth symmetric power of the conormal bundle of $\mathbf{A}^I$ and there are exact sequences

$$0 \to \overset{-k}{×}\overset{0}{———}•\overset{-k}{———}× \to \mathcal{O}_{(k)} \to \mathcal{O}_{(k-1)} \to 0. \tag{21}$$

An element of $H^1(\mathbf{A}^I, \mathcal{O})$ extends to an element of $H^1(\mathbf{P}^I \times \mathbf{P}^{*I}, \mathcal{O})$ only if it extends to each $H^1(\mathbf{A}^I, \mathcal{O}_{(k)})$. This yields an infinite series of extension problems which can readily be solved using the Penrose transform.

We first attempt extension to $H^1(\mathbf{A}^I, \mathcal{O}_{(1)})$. The long exact sequence in cohomology, applied to the sequence (21) with $k = 1$, and the Penrose transform yield the exactness of

$$0 \to \Gamma(\mathbf{M}^I, •\overset{0}{——}\overset{-2}{×}\overset{0}{——}•) \to H^1(\mathbf{A}^I, \mathcal{O}_{(1)}) \to H^1(\mathbf{A}^I, \mathcal{O}) \to 0.$$

Each of the spaces in this sequence are modules over $\mathbf{sl}(4, \mathbf{C})$; the outer two have distinct infinitesimal character (see chapter 11) and so this sequence is split. It follows that every element of $H^1(\mathbf{A}^I, \mathcal{O})$ lifts uniquely to an element of $H^1(\mathbf{A}^I, \mathcal{O}_{(1)})$.

Next, we attempt extension to $H^1(\mathbf{A}^I, \mathcal{O}_{(2)})$. Sequence (21), with $k = 2$, and the Penrose transform yield the following exact sequence of $\mathbf{sl}(4, \mathbf{C})$-modules

$$0 \to H^1(\mathbf{A}^I, \mathcal{O}_{(2)}) \to H^1(\mathbf{A}^I, \mathcal{O}_{(1)}) \overset{\phi_2}{\to} \Gamma(\mathbf{M}^I, •\overset{0}{——}\overset{-2}{×}\overset{0}{——}•).$$

Infinitesimal character implies that the image under ϕ_2 of any element lifted from $H^1(\mathbf{A}^I, \mathcal{O})$ must be zero and, canonically,

$$H^1(\mathbf{A}^I, \mathcal{O}) \cong H^1(\mathbf{A}^I, \mathcal{O}_{(2)}).$$

Lastly, we attempt extension to $H^1(\mathbf{A}^I, \mathcal{O}_{(3)})$. We obtain the exact sequence

$$0 \to H^1(\mathbf{A}^I, \mathcal{O}_{(3)}) \to H^1(\mathbf{A}^I, \mathcal{O}_{(2)}) \overset{\phi_3}{\to} \Gamma(\mathbf{M}^I, \Omega^3) \overset{d}{\to} \Gamma(\mathbf{M}^I, \Omega^4)$$

using the Penrose transform to identify $H^2(\mathbf{A}^I, ×\overset{-3}{——}•\overset{0}{——}×\overset{-3}{})$. Then ϕ_3 corresponds to $d_* : \Gamma(\mathbf{M}^I, \Omega^2) \to \Gamma(\mathbf{M}^I, \Omega^3)$, and $H^1(\mathbf{A}^I, \mathcal{O}_{(3)})$ is identified with Maxwell fields on $\mathbf{M}^I$.

The obstructions to further liftings lie in $H^2(\mathbf{A}^I, \overset{\text{-k}\ \ 0\ \ \text{-k}}{\times\!\!-\!\!\bullet\!\!-\!\!\times})$ for $k \geq 4$; the Penrose transform identifies these spaces with the kernels of appropriate differential operators on $\mathbf{M}^I$. It is easy to see that these spaces have an infinitesimal character distinct from that of $H^1(\mathbf{A}^I, \mathcal{O}_{(3)})$ so that all further obstructions vanish. In fact, with a little further work, it can be shown that

$$H^1(\mathbf{A}^I, \mathcal{O}_{(3)}) \cong H^1(\mathbf{P}^I \times \mathbf{P}^{*I}, \mathcal{O}).$$

Remark (9.5.1).   These results may also be deduced from a *supersymmetric* correspondence and its Penrose transform [52,62,161].

Remark (9.5.2).   A very similar construction exists for

$$H^1(\mathbf{A}^I, \overset{1\ \ 0\ \ 1}{\times\!\!-\!\!\bullet\!\!-\!\!\times}) \cong \frac{\overset{2\ \ \text{-2}\ \ 2}{\bullet\!\!-\!\!\times\!\!-\!\!\bullet}}{\text{im}\ \overset{1\ \ 0\ \ 1}{\bullet\!\!-\!\!\times\!\!-\!\!\bullet} \rightarrow \overset{2\ \ \text{-2}\ \ 2}{\bullet\!\!-\!\!\times\!\!-\!\!\bullet}}$$

$$\cong \ker \Gamma(\mathbf{M}^I, \overset{4\ \ \text{-4}\ \ 0}{\bullet\!\!-\!\!\times\!\!-\!\!\bullet} \oplus \overset{0\ \ \text{-4}\ \ 4}{\bullet\!\!-\!\!\times\!\!-\!\!\bullet}) \rightarrow \Gamma(\mathbf{M}^I, \overset{2\ \ \text{-6}\ \ 2}{\bullet\!\!-\!\!\times\!\!-\!\!\bullet})$$

which corresponds to infinitesimal conformal deformations of $\mathbf{M}^I$ (and to deformations of $\mathbf{A}^I$ which preserve its *contact structure*, viz. $\Theta_{\mathbf{A}} \rightarrow \overset{1\ \ 0\ \ 1}{\times\!\!-\!\!\bullet\!\!-\!\!\times}$ [12]). We find that

$$H^1(\mathbf{A}^I, \overset{1\ \ 0\ \ 1}{\times\!\!-\!\!\bullet\!\!-\!\!\times}) \cong H^1(\mathbf{A}^I, \mathcal{O}_{(4)} \otimes \overset{1\ \ 0\ \ 1}{\times\!\!-\!\!\bullet\!\!-\!\!\times})$$

and

$$0 \rightarrow H^1(\mathbf{A}^I, \mathcal{O}_{(5)} \otimes \overset{1\ \ 0\ \ 1}{\times\!\!-\!\!\bullet\!\!-\!\!\times}) \rightarrow H^1(\mathbf{A}^I, \mathcal{O}_{(4)} \otimes \overset{1\ \ 0\ \ 1}{\times\!\!-\!\!\bullet\!\!-\!\!\times})$$

$$\overset{\varphi}{\rightarrow} \Gamma(\mathbf{M}^I, \overset{2\ \ \text{-6}\ \ 2}{\bullet\!\!-\!\!\times\!\!-\!\!\bullet}) \rightarrow \Gamma(\mathbf{M}^I, \overset{1\ \ \text{-6}\ \ 1}{\bullet\!\!-\!\!\times\!\!-\!\!\bullet}) \rightarrow 0.$$

Hence $H^1(\mathbf{A}^I, \mathcal{O}_{(5)} \otimes \overset{1\ \ 0\ \ 1}{\times\!\!-\!\!\bullet\!\!-\!\!\times})$ corresponds to the space of spinors fields $\tilde{\psi}_{A'B'C'D'}$ and ψ_{ABCD} on $\mathbf{M}^I$, satisfying

$$\{\nabla_A^{C'} \nabla_B^{D'} + \Phi_{AB}^{C'D'}\}\tilde{\psi}_{A'B'C'D'} = \{\nabla_{A'}^C \nabla_{B'}^D + \Phi_{A'B'}^{CD}\}\psi_{ABCD} = \tfrac{1}{2}B_{ab} = 0$$

where, again, ∇ is allowed to be the Levi–Civita connection of any metric in the conformal class on $\mathbf{M}^I$.

B_{ab} is a linearized form of the *Bach* tensor, defined by the same formula on an arbitrary four dimensional conformal manifold $\mathcal{M}$. This is important, for over such a manifold the vanishing of the Bach tensor is a necessary condition for the existence of an Einstein metric in the conformal class [103]. $\mathbf{A}^I$ may be replaced by the space $\mathcal{A}$ of null geodesics in $\mathcal{M}$ although analogues of $\mathbf{P}$ and $\mathbf{P}^*$ do not exist. Nonetheless, it seems that

$$H^1(\mathbf{A}^I, \mathcal{O}_{(5)} \otimes \overset{\displaystyle 1\quad 0\quad 1}{\times\!\!-\!\!\bullet\!\!-\!\!\times})$$

corresponds to deformations of $\mathcal{O}_{(5)}$ preserving some kind of contact structure, and that $\mathcal{A}$ might possess a formal extension to order five if, and only if, $\mathcal{M}$ has vanishing Bach tensor [12,59,107].

9.6   Higher dimensions—conformal case

It is now time to employ our machinery to compute the Penrose transform for higher dimensional spaces which generalize the standard Minkowski space of four dimensions. There are really two ways to go. One is to focus on conformal structure, arguing that physically the relevant structure in four dimensions is the metric or conformal structure in Minkowski space and that in higher dimensions this ought to be available. The second is to concentrate on the isomorphism

$$\overset{\displaystyle 1\quad 0\quad 1}{\bullet\!\!-\!\!\times\!\!-\!\!\bullet} = \overset{\displaystyle 0\quad 0\quad 1}{\bullet\!\!-\!\!\times\!\!-\!\!\bullet} \otimes \overset{\displaystyle 1\quad 0\quad 0}{\bullet\!\!-\!\!\times\!\!-\!\!\bullet}$$

or, in abstract index notation,

$$\mathcal{O}^a = \mathcal{O}^{AA'} = \mathcal{O}^A \otimes \mathcal{O}^{A'}$$

and seek higher dimensional spaces on which the tangent bundle factors similarly. The first case evidently means considering higher dimensional complex quadrics (complexified spheres)—we take up this case in this section and return to it in greater detail when we consider non-standard homomorphisms below. The second case concerns Grassmannians and is the subject of the next section.

Consider the general complex quadric in $\mathbf{CP}^S$; this bears a conformal structure which is easy to specify geometrically—two points are null separated if the straight line joining them lies entirely in the quadric. There is a distinction between odd and even dimensional cases because of the distinction between odd and even orthogonal Lie algebras so these will be taken separately. It is, in fact, quite remarkable how large this distinction is under the Penrose transform. An example of this is the fact that whilst in even dimensions there is an analogue of the conformally invariant Laplacian of four dimensions no such operator exists in odd dimensions (unless we restrict ourselves to *real* manifolds; the difficulty is that such a Laplacian would need to be defined on fractionally conformally weighted functions [56]).

Zero rest mass fields in even dimensions

We shall continue to denote an affine open "big cell" in $\mathbf{CS}^{2n}$ by $\mathbf{M}^I$. The corresponding open subvariety of the variety of pure (reduced) spinors of one kind will be denoted by $\mathbf{P}^I$. This corresponds to $\mathbf{M}^I$ under the double fibration

where each diagram has $n+1$ nodes. The fibres of η are n-dimensional projective spaces whilst those of τ are isomorphic with the twistor space $\mathbf{Z}^{2n-2}$ two dimensions down. Recall that $\mathbf{Z}^{2n}$ is $n(n+1)/2$ dimensional. The analogue of the Hopf line bundle and its powers on projective spaces must be

since these are the only homogeneous line bundles on $\mathbf{Z}^{2n}$.

When we return to the Penrose transform for quadrics to compute homomorphisms of Verma modules, we shall compute rather more directly using the group structure of the Weyl group in detail. For the moment, we will be content to compute explicitly in the Dynkin diagram notation. The Bernstein–Gelfand–Gelfand resolution of the inverse image of $\mathcal{O}(k)$ is

$$0 \to \eta^{-1}\mathcal{O}(k) \to \Delta_\eta^\bullet$$

where

$$\Delta_\eta^{n-1} = \quad \overset{-n}{\times}\!\!-\!\!\overset{0}{\bullet}\!\!-\!\!\overset{0}{\bullet}\cdots\overset{0}{\bullet}\!\!-\!\!\overset{0}{\bullet}\!\!\underset{\bullet\,1}{\overset{\times\,k+1}{<}}$$

$$\Delta_\eta^{n} = \quad \overset{-n\text{-}1}{\times}\!\!-\!\!\overset{0}{\bullet}\!\!-\!\!\overset{0}{\bullet}\cdots\overset{0}{\bullet}\!\!-\!\!\overset{0}{\bullet}\!\!\underset{\bullet\,0}{\overset{\times\,k+2}{<}}$$

(which is the relative deRham resolution). The direct images of these resolvents vary with k and n, much as in the four dimensional case. Again, the singular and non-singular cases behave quite distinctly. The non-singular case has two subcases (obtaining a "fields" or "potential modulo gauges" description), corresponding to $p \le -2n$ and $p \ge 0$. For $p \le -2n$ the unique dominant weight conjugate under the affine action of $W_{\mathbf{g}}$ to the weights in the above BGG resolution is

$$\overset{0}{\bullet}\!\!-\!\!\overset{0}{\bullet}\!\!-\!\!\overset{0}{\bullet}\cdots\overset{0}{\bullet}\!\!-\!\!\overset{0}{\bullet}\!\!\underset{\bullet\,0}{\overset{\bullet\,-k\text{-}2n}{<}} \qquad \text{if } n \text{ is odd}$$

or

$$\overset{0}{\bullet}\!\!-\!\!\overset{0}{\bullet}\!\!-\!\!\overset{0}{\bullet}\cdots\overset{0}{\bullet}\!\!-\!\!\overset{0}{\bullet}\!\!\underset{\bullet\,-k\text{-}2n}{\overset{\bullet\,0}{<}} \qquad \text{if } n \text{ is even.}$$

In both cases, the direct images of the resolvents all occur in the same degree; therefore, the hypercohomology spectral sequences collapse to a single complex in one row. This complex is a subcomplex of the BGG resolution for the conjugate dominant weight on $\mathbf{M}$. Below we give the relevant BGG resolutions and encircle where the direct images fall. In all cases, the subcomplex begins or ends in the middle degree of the BGG resolution.

Non-singular case: $k \ge 0$: negative helicity fields

In this case, all resolvents have non-trivial direct images only in degree zero, yielding the subcomplex encircled:

$$\underset{\bullet\,0}{\overset{0\,0\,0\quad 0\,0\,\bullet\,k}{\times\!\bullet\!\bullet\cdots\bullet\!\bullet}} \to \cdots \to \underset{\bullet\,1}{\overset{-n\,0\,0\quad 0\,0\,\bullet\,k+1}{\times\!\bullet\!\bullet\cdots\bullet\!\bullet}} \to \underset{\bullet\,0}{\overset{-n\text{-}1\,0\,0\quad 0\,0\,\bullet\,k+2}{\times\!\bullet\!\bullet\cdots\bullet\!\bullet}}$$

$$\oplus$$

$$\underset{\bullet\,k+2}{\overset{-k\text{-}n\text{-}1\,0\,0\quad 0\,0\,\bullet\,0}{\times\!\bullet\!\bullet\cdots\bullet\!\bullet}} \to \underset{\bullet\,k+1}{\overset{-k\text{-}n\text{-}2\,0\,0\quad 0\,0\,\bullet\,1}{\times\!\bullet\!\bullet\cdots\bullet\!\bullet}}$$

It follows that

$$H^{n-1}(\mathbf{P}^I, \mathcal{O}(k)) \;\cong\; \ker \quad \underset{k+2}{\overset{-k-n-1 \;\; 0 \;\;\; 0 \qquad 0 \;\; 0 \quad 0}{\times\!-\!\bullet\!-\!\bullet\cdots\bullet\!-\!\bullet\!\!\big\langle}} \quad \overset{\eth}{\to} \quad \underset{k+1}{\overset{-k-n-2 \;\; 0 \;\;\; 0 \qquad 0 \;\; 0 \quad 1}{\times\!-\!\bullet\!-\!\bullet\cdots\bullet\!-\!\bullet\!\!\big\langle}}$$

using the exactness at the central square of the BGG resolution. Recall (3.1.8) the definition of spinor bundles on $\mathbf{CS}^{2n}$. Then the subject of $\eth$ is the highest irreducible component of the symmetric spinor bundle $\mathcal{O}^{(\alpha_1\ldots\alpha_{k+2})}[-k-n-1]$. Its image is the highest irreducible component of $\mathcal{O}^{(\alpha_1\ldots\alpha_{k+1})\alpha'}[-k-n-2]$. Furthermore, $\eth$ is a first-order differential operator (as is readily checked by observing that the only possible symbol is of first order). It is obtained by applying the *Dirac operator* [127] and projecting out the highest irreducible component of the result (which is the only part of the Dirac operator in higher dimensions that is conformally invariant). It is reasonable, therefore, to identify $H^{n-1}(\mathbf{P}^I, \mathcal{O}(k))$ with a higher dimensional analogue of the *negative helicity zero rest mass free fields* of four dimensions.

Non-singular case : $k \leq -2n$

As should be expected, $H^*(\mathbf{P}^I, \mathcal{O}(k))$ again corresponds to an analogue of the zero-rest-mass fields; a novel feature is that the helicity of the result depends on the parity of n.

All resolvents have non-trivial direct images only in degree $\frac{n(n-1)}{2}$ so that the hypercohomology spectral sequence collapses to a subcomplex of a BGG resolution on $\mathbf{M}^{2n}$, encircled in the following diagrams:

n even : positive helicity fields

so that

$$H^{\frac{n(n-1)}{2}}(\mathbf{P}^I, \mathcal{O}(k)) \;\cong\; \ker \quad \underset{0}{\overset{k+n-1 \;\; 0 \;\;\; 0 \qquad 0 \;\; 0 \quad -k-2n+2}{\times\!-\!\bullet\!-\!\bullet\cdots\bullet\!-\!\bullet\!\!\big\langle}} \quad \overset{\eth'}{\to} \quad \underset{1}{\overset{k+n-2 \;\; 0 \;\;\; 0 \qquad 0 \;\; 0 \quad -k-2n+1}{\times\!-\!\bullet\!-\!\bullet\cdots\bullet\!-\!\bullet\!\!\big\langle}}$$

$\eth'$ acts on the highest irreducible component of $\mathcal{O}^{(\alpha_1\ldots\alpha_{-k-2n+2})}[k+n-1]$ and so the cohomology group may be identified as an extension of positive helicity zero rest mass fields.

n odd: negative helicity fields

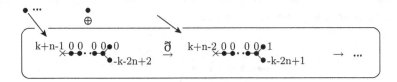

so that

$$H^{\frac{n(n-1)}{2}}(\mathbf{P}^I, \mathcal{O}(k)) \cong \ker \quad \overset{k+n-1\;\;0\;\;0\;\;\;\;\;0\;\;0\quad 0}{\underset{-k-2n+2}{\times\!\!-\!\!\bullet\!\!-\!\!\bullet\cdots\bullet\!-\!\bullet}} \quad \overset{\eth}{\longrightarrow} \quad \overset{k+n-2\;\;0\;\;0\;\;\;\;\;0\;\;0\quad 1}{\underset{-k-2n+1}{\times\!\!-\!\!\bullet\!\!-\!\!\bullet\cdots\bullet\!-\!\bullet}}$$

that is negative helicity zero rest mass free fields, again. Observe that when *n* is odd *only* negative helicity fields occur in the cohomology of the sheaves $\mathcal{O}(k)$, irrespective of the sign of *k*. To obtain positive frequency fields it is necessary to compute cohomology on the other space of pure spinors for $\mathbf{so}(2n+2,\mathbf{C})$. Notice that it is precisely in this case that the two spin spaces are *not* contragredient representations; naïvely, they should not be expected to hold the same "information content".

Singular cases $k = -2n+1$ or -1

When $k = -2n+1$, only the zeroth and first terms of the BGG resolution have non-trivial direct images:

$$\tau_*^{\frac{n(n-1)}{2}} \quad \overset{0\;\;0\;\;0\;\;\;\cdots\;\;0\;\;0}{\underset{0}{\times\!\!-\!\!\bullet\!\!-\!\!\bullet\cdots\bullet\!\!-\!\!\bullet}}\!\!\overset{-2n+1}{\diagdown} = \begin{cases} \mathcal{O}^{\alpha}[-n] & n \text{ odd} \\ \mathcal{O}^{\alpha'}[-n] & n \text{ even} \end{cases}$$

$$\tau_*^{\frac{n(n-1)}{2}} \quad \overset{-2\;\;1\;\;0\;\;\;\cdots\;\;0\;\;0}{\underset{0}{\times\!\!-\!\!\bullet\!\!-\!\!\bullet\cdots\bullet\!\!-\!\!\bullet}}\!\!\overset{-2n+1}{\diagdown} = \begin{cases} \mathcal{O}^{\alpha'}[-n-1] & n \text{ odd} \\ \mathcal{O}^{\alpha}[-n-1] & n \text{ even}. \end{cases}$$

Therefore, $H^{\frac{n(n-1)}{2}}(\mathbf{P}^I, \mathcal{O}(k))$ consists of helicity $\pm\frac{1}{2}$ zero rest mass fields, depending on the parity of *n*. The reader will easily check that for $k = -1$, helicity $\pm\frac{1}{2}$ fields are obtained.

Singular cases $k = -2n + 2$ or -2

In the case $k = -2n + 2$, only the zeroth and second terms of the BGG resolution have non-trivial direct images:

$$\tau_* \overset{\frac{n(n-1)}{2}}{} \quad \underset{\times}{\overset{0}{\bullet}}\,\underset{}{\overset{0}{\bullet}}\,\underset{}{\overset{0}{\bullet}} \cdots \underset{}{\overset{0}{\bullet}}\,\underset{}{\overset{0}{\bullet}}\,\overset{-2n+2}{\underset{0}{\times}} \; = \mathcal{O}[-n+1]$$

$$\tau_* \overset{\frac{n(n-1)}{2}-1}{} \quad \underset{\times}{\overset{-1}{\bullet}}\,\underset{}{\overset{0}{\bullet}}\,\underset{}{\overset{1}{\bullet}} \cdots \underset{}{\overset{0}{\bullet}}\,\underset{}{\overset{0}{\bullet}}\,\overset{-2n+2}{\underset{0}{\times}} \; = \mathcal{O}[-n-1]$$

and the hypercohomology spectral sequence is

$$E_1^{p,q} = E_2^{p,q} \left| \begin{array}{cccc} \mathcal{O}[-n+1] & 0 & 0 & \cdots \\ 0 & 0 & \mathcal{O}[-n-1] & \cdots \\ \vdots & \vdots & \vdots & \end{array} \right.$$

so that

$$H^{\frac{n(n-1)}{2}}(\mathbf{P}^I, \mathcal{O}(-2n+2)) \cong \ker \square : \mathcal{O}[-n+1] \to \mathcal{O}[-n-1]$$

where $\square$ is the Laplacian. The reader may check that

$$H^{n-1}(\mathbf{P}^I, \mathcal{O}(-2)) \cong \ker \square : \mathcal{O}[-n+1] \to \mathcal{O}[-n-1].$$

$\square$ is, again, a *non-standard* differential operator.

Other singular characters

For the remaining singular cases, $-2n + 2 < k < -2$, it turns out that all cohomology vanishes:

Theorem (9.6.1). *If $-2n + 2 < k < -2$, then*

$$H^*\left(\underset{}{\overset{0}{\bullet}}\,\underset{}{\overset{0}{\bullet}}\,\underset{}{\overset{0}{\bullet}} \cdots \underset{}{\overset{0}{\bullet}}\,\underset{}{\overset{0}{\bullet}}\,\overset{k}{\underset{0}{\times}} \right) = 0.$$

The cases $n = 3, 4$ are easily checked directly—the remainder follow by induction.

Zero rest mass fields in odd dimensions

The non-singular cases in odd dimensions are not very different from those in even dimensions; the singular cases are. This is because of the lack of non-standard differential operators in odd dimensions.

The double fibration is

where each diagram has $n+1$ nodes. $\mathbf{M}^I$ and $\mathbf{P}^I$ will denote an affine $2n+1$ dimensional Minkowski space and the corresponding twistor space.

Notice that ●———●———● ⋯ ●⇒⟩× is the same as ●———●———● ⋯ ●⟨

($n+2$ nodes) as a manifold (but not as a homogeneous space [128])—in particular, $\mathbf{Z}^{2n+1}$ has dimension $\frac{(n+1)(n+2)}{2}$.

The Hasse graph for the η–fibres is

$$W_{\mathbf{r}}^{\mathbf{q}} = \{id, \sigma_1, \sigma_1\sigma_2, \ldots, r_j = \sigma_1\sigma_2 \cdots \sigma_j, \ldots, r_n\}$$

and so, if we let

$$\mathcal{O}(k) = \overset{0\quad 0\quad 0\qquad 0\quad k}{●———●———● \cdots ●⇒⟩×}$$

then the ith term in the BGG resolution of $\eta^{-1}\mathcal{O}(k)$ is

$$\Delta_\eta^0 = \overset{0\quad 0\quad 0\qquad 0\quad k}{×———●———● \cdots ●⇒⟩×}$$

$$\Delta_\eta^i = \overset{\text{-i-1}\ 0\quad 0\qquad 0\ 1\ 0\qquad 0\quad k}{×———●———● \cdots ●———●———● \cdots ●⇒⟩×} \qquad \text{for } 1 \leq i \leq n-1$$
$$\underset{\text{node } i+1 \uparrow}{}$$

$$\Delta_\eta^n = \overset{\text{-n-1}\ 0\quad 0\qquad 0\quad k+1}{×———●———● \cdots ●⇒⟩×} \ .$$

If $k \geq 0$, then the zeroth row of the first level in the hypercohomology spectral sequence is simply the subcomplex of the Bernstein–Gelfand–Gelfand resolution of

$$\lambda = \overset{0\quad 0\quad 0\qquad 0\quad k}{●———●———● \cdots ●⇒⟩×} \quad \text{on} \quad ×———●———● \cdots ●⇒—●$$

consisting of the first n resolvents (since $W_{\mathbf{r}}^{\mathbf{q}} \subset W^{\mathbf{p}}$). If $r_i \in W^{\mathbf{p}}$ (for
×—•—• · · · •⟹•) denotes the unique element of length i (see example 4.3.8), then the exactness of this BGG resolution implies

$$H^n(\mathbf{P}^I, \mathcal{O}(k)) \cong \text{coker } \mathcal{O}_{\mathbf{p}}(r_{n-1}.\lambda) \xrightarrow{\not\partial} \mathcal{O}_{\mathbf{p}}(r_n.\lambda)$$

$$\cong \ker \mathcal{O}_{\mathbf{p}}(r_{n+1}.\lambda) \xrightarrow{\not\partial} \mathcal{O}_{\mathbf{p}}(r_{n+2}.\lambda)$$

$$\cong \ker \quad \overset{-k-n-2 \; 0 \quad 0 \qquad 0 \; k+1}{\text{×—•—• · · · •⟹•}} \;\longrightarrow\; \overset{-n-k-3 \; 0 \quad 0 \qquad 1 \quad k}{\text{×—•—• · · · •⟹•}}$$

and cohomology in all other positive degrees vanishes. $\not\partial$ is again the conformally invariant part of a Dirac operator, and a space of zero rest mass fields is obtained. (In odd dimensions, the concept of left- or right-handedness does not occur since there is only one irreducible spinor representation).

If $k \leq -2n - 1$, then the BGG resolution consists of bundles which have nontrivial τ-direct images only in degree $n(n+1)/2$ (which is, of course, the τ-fibre dimension). The result is a first term in the hypercohomology spectral sequence whose $n(n+1)/2^{\text{th}}$ row is part of the BGG resolution on $\mathbf{M}$ just obtained so that

$$H^{\frac{n(n+1)}{2}}(\mathbf{P}^I, \mathcal{O}(k)) \cong \ker \mathcal{O}_{\mathbf{p}}(r_{n+1}.\lambda) \xrightarrow{\not\partial} \mathcal{O}_{\mathbf{p}}(r_{n+2}.\lambda)$$

as before.

This deals with the regular cases. We will leave the general singular case to the reader as an exercise; observe, typically, that in five dimensions, the hypercohomology sequence in the Penrose transform of $\mathcal{O}(-2)$ is

$$E_1^{p,q} \cong E_\infty^{p,q} \begin{array}{|ccc} 0 & \overset{-2 \; 0 \quad 0}{\text{×—•⟹•}} & 0 \\ \hline 0 & 0 & \overset{-3 \quad 0 \quad 0}{\text{×—•⟹•}} \end{array}$$

which is rather different from the corresponding even dimensional case! This is closely related to the lack of non-standard homomorphisms of integral Verma modules induced from $\mathbf{p}$ in this case (see [56]).

9.7   A Grassmannian generalization

An alternative higher dimensional analogue of Penrose's four dimensional twistor theory is obtained by generalizing the factorization of the tangent bundle of Minkowski space into a tensor product of vector bundles:

$$\mathcal{O}^a = \mathcal{O}^A \otimes \mathcal{O}^{A'}$$

or, in our Dynkin diagram notation,

$$
\begin{array}{ccc}
\underset{\bullet\!-\!\times\!-\!\bullet}{1\ \ 0\ \ 1} & = & \underset{\bullet\!-\!\times\!-\!\bullet}{1\ \ 0\ \ 0} \otimes \underset{\bullet\!-\!\times\!-\!\bullet}{0\ \ 0\ \ 1}\ .
\end{array}
$$

A quick glance at table 3.2 shows that the tangent bundle of a generalized flag variety is irreducible and factors like this only if the variety is a Grassmannian

$$
\mathbf{Gr}_p(\mathbf{C}^{p+q}) = \underset{\bullet\!-\!\bullet\ \cdots\ \bullet\!-\!\times\!-\!\bullet\ \cdots\ \bullet\!-\!\bullet}{\qquad\qquad\quad p^{th}\ \text{node}}
$$

(with $p+q-1$ nodes in all; $p,q \geq 2$ to avoid degeneracy). Then its tangent bundle is

$$
\begin{aligned}
\Theta &= \underset{\bullet\!-\!\bullet}{1\ \ 0}\ \cdots\ \underset{\bullet\!-\!\times\!-\!\bullet}{0\ \ 0\ \ 0}\ \cdots\ \underset{\bullet\!-\!\bullet}{0\ \ 1} \\
&= \underset{\bullet\!-\!\bullet}{1\ \ 0}\ \cdots\ \underset{\bullet\!-\!\times\!-\!\bullet}{0\ \ 0\ \ 0}\ \cdots\ \underset{\bullet\!-\!\bullet}{0\ \ 0} \otimes \underset{\bullet\!-\!\bullet}{0\ \ 0}\ \cdots\ \underset{\bullet\!-\!\times\!-\!\bullet}{0\ \ 0\ \ 0}\ \cdots\ \underset{\bullet\!-\!\bullet}{0\ \ 1}\ .
\end{aligned}
$$

In Penrose's abstract index notation we should denote

$$
\begin{aligned}
\mathcal{O}^{A'} &= \underset{\bullet\!-\!\bullet}{1\ \ 0}\ \cdots\ \underset{\bullet\!-\!\times\!-\!\bullet}{0\ \ 0\ \ 0}\ \cdots\ \underset{\bullet\!-\!\bullet}{0\ \ 0} \\
\mathcal{O}^{A} &= \underset{\bullet\!-\!\bullet}{0\ \ 0}\ \cdots\ \underset{\bullet\!-\!\times\!-\!\bullet}{0\ \ 0\ \ 0}\ \cdots\ \underset{\bullet\!-\!\bullet}{0\ \ 1}
\end{aligned}
$$

(which are bundles of ranks p and q respectively) so that

$$
\mathcal{O}^a = \Theta = \mathcal{O}^{AA'}\ .
$$

With this settled, we must now decide on an accompanying twistor space. There are two strong contenders: the first is obtained by recalling that in the four dimensional case a point in twistor space corresponds to a plane in Minkowksi space. A moment's thought shows that if we wish to generalize this fact then we should choose to replace twistor space by $\mathbf{Gr}_{p-1}(\mathbf{C}^{p+q})$ and consider the following double fibration:

(22)

For then the fibres of η are $\mathbf{CP}^q$'s and project to the desired planes. (Using $p+1$ instead of $p-1$, yields the analogue of dual twistor space.) This option seems best from a physical point of view because it will admit a non-linear graviton-like construction—see [6,123] for details.

A second possibility is afforded by

$$(23)$$

This is an interesting choice from the representation theory point of view for, as we shall see, it leads to the ladder representations of $SU(p,q)$ [47,54].

In both cases there is a difference depending on whether $p \geq 3$ or not. (If $p = 2$ the cases agree on twistor spaces, if not on their duals.) We consider $p, q \geq 3$ and leave the remaining cases for the reader (equipped with fresh pencils and paper!). Let X denote an affine or Stein region of $\mathbf{Gr}_p(\mathbf{C}^{p+q})$ and set $Y = \tau^{-1}X$, $Z = \eta Y$, as usual.

Consider (22). Let λ be dominant for $\mathbf{g} = sl(p+q, \mathbf{C})$. For example we might take $\lambda = \lambda_k$ where

$$\lambda_k = \overset{0\quad 0\quad\quad -p+k\ 0\quad 0\quad\quad 0\quad 0}{\bullet\!\!-\!\!\bullet\ \cdots\ \times\!\!-\!\!\bullet\!\!-\!\!\bullet\ \cdots\ \bullet\!\!-\!\!\bullet}$$

with $k \geq p$ and then $\mathcal{O}(-p+k) = \mathcal{O}_{\mathbf{r}}(\lambda_k)$ is a line bundle on Z and its sections are a natural substitute for the ordinary homogeneous twistor functions.

Our aim is to compute

$$H^i(Z, \mathcal{O}_{\mathbf{r}}(\lambda))$$

via the Penrose transform. Using the methods of section 4.4 we quickly compute that the Hasse diagram associated to the fibres of η is

$$W_{\mathbf{r}}^{\mathbf{q}} = \{id, \sigma_p, \ldots, \Sigma_k = \sigma_p \cdots \sigma_{p+k-1}, \ldots \Sigma_q\}$$

so that the ith resolvent in the BGG resolution of $\eta^{-1}\mathcal{O}_{\mathbf{r}}(\lambda)$ is

$$\Delta_\eta^i = \mathcal{O}_{\mathbf{q}}(\Sigma_i.\lambda).$$

Because λ is dominant, these only have zeroth direct images:

$$\mathcal{R}^i = \tau_* \Delta_\eta^i = \mathcal{O}_{\mathbf{p}}(\Sigma_i.\lambda).$$

$\mathcal{R}^\bullet$ is a subcomplex of the BGG resolution of $E(\lambda)$ on X, and its cohomology is $H^*(Z, \mathcal{O}_{\mathbf{r}}(\lambda))$, by the Penrose transform. In particular, we have

$$H^1(Z, \mathcal{O}_{\mathbf{r}}(\lambda)) \cong \frac{\ker\ \mathcal{O}_{\mathbf{p}}(\Sigma_1.\lambda) \to \mathcal{O}_{\mathbf{p}}(\Sigma_2.\lambda)}{\operatorname{im}\ \mathcal{O}_{\mathbf{p}}(\lambda) \to \mathcal{O}_{\mathbf{p}}(\Sigma_1.\lambda)}$$

so that

$$H^1(Z,\mathcal{O}(-p+k))$$

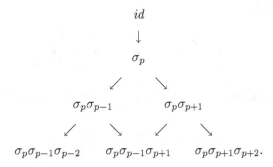

This is in *potential modulo gauge* form as we should expect. To obtain a field formulation we must use the exactness of the BGG resolution on X. The first few terms of this are determined by the following terms of $W^{\mathbf{p}}$:

$$
\begin{array}{c}
id \\
\downarrow \\
\sigma_p \\
\swarrow \qquad \searrow \\
\sigma_p\sigma_{p-1} \qquad\qquad \sigma_p\sigma_{p+1} \\
\swarrow \qquad \searrow \quad \swarrow \qquad \searrow \\
\sigma_p\sigma_{p-1}\sigma_{p-2} \quad \sigma_p\sigma_{p-1}\sigma_{p+1} \quad \sigma_p\sigma_{p+1}\sigma_{p+2}.
\end{array}
$$

(This is the first point at which the case $p = 2$ differs from the others; it has no "$\sigma_p\sigma_{p-1}\sigma_{p-2}$".) Thus

$$H^1(Z,\mathcal{O}_{\mathbf{r}}(\lambda)) \cong \ker d : \mathcal{O}_{\mathbf{p}}(\sigma_p\sigma_{p-1}.\lambda) \to \begin{array}{c} \mathcal{O}_{\mathbf{p}}(\sigma_p\sigma_{p-1}\sigma_{p-2}.\lambda) \\ \oplus \\ \mathcal{O}_{\mathbf{p}}(\sigma_p\sigma_{p-1}\sigma_{p+1}.\lambda) \end{array}$$

(there being only one irreducible term in the range of d when $p = 2$).

When $\lambda = 0$, since the tangent bundle of a Grassmanian is irreducible as a homogeneous bundle, the BGG resolution is just the de Rham resolution of $\mathbf{C}$. Then d is just the exterior differential applied to an irreducible summand of the two-forms. In abstract index notation,

$$\Omega^2 = \mathcal{O}_{\mathbf{p}}(\sigma_p\sigma_{p-1}.0) \oplus \mathcal{O}_{\mathbf{p}}(\sigma_p\sigma_{p+1}.0)$$
$$= \mathcal{O}_{[A'B'](AB)} \oplus \mathcal{O}_{(A'B')[AB]}$$

and d is

$$\phi_{[A'B'](AB)} \to \nabla_{(A|[A'}\phi_{B'C']|BC)} \oplus \nabla_{[A|(A'}\phi_{B')C'|B]C}$$

(the first term, evidently, being absent if $p = 2$). The connections ∇ on $\mathcal{O}^A$ and $\mathcal{O}^{A'}$ are determined by requiring that they agree on the determinant bundle

$$\epsilon \in \quad \overset{0 \quad 0}{\bullet\!\!-\!\!\bullet} \cdots \overset{0 \quad 1 \quad 0}{\bullet\!\!-\!\!\times\!\!-\!\!\bullet} \cdots \overset{0 \quad 0}{\bullet\!\!-\!\!\bullet}$$

where they preserve a section ϵ and that the induced connection on Θ be torsion free [6]. The operators in the Penrose transform are all invariant of ϵ.

We might wish to interpret the kernel of d as left-handed electromagnetism and, by analogy, the remaining $H^1(Z, \mathcal{O}(k))$, $k \geq 0$, as left-handed generalizations of massless fields. It is not clear what the physical nature of such fields might be, however.

Now consider $H^*(Z, \mathcal{O}(-p+k))$ with general k. We find that the jth BGG resolvent is

$$\overset{0 \quad\quad -p+k+j\text{-}j\text{-}1 \quad 0 \quad\quad\quad 1 \quad\quad\quad 0}{\bullet \cdots \times\!\!-\!\!\times\!\!-\!\!\bullet \cdots \bullet \cdots \bullet} \atop \qquad\qquad\qquad\qquad\qquad\uparrow\; p+j^{\text{th}}\text{ node}$$

If $j \leq -k$ this has only a non-trivial $p-1^{\text{st}}$ direct image of

$$\overset{\text{-k-j} \quad\quad 0 \quad \text{-p+k} \quad 0 \quad\quad 1 \quad\quad\quad 0}{\bullet \cdots \bullet\!\!-\!\!\times\!\!-\!\!\bullet \cdots \bullet \cdots \bullet} \atop \qquad\qquad\qquad\qquad\qquad\uparrow\; p+j^{\text{th}}\text{ node}$$

If $-k+1 \leq j \leq p-k-1$ then there are *no* non-zero direct images whilst if $j \geq p-k$ only the zeroth direct image is non-trivial:

$$\overset{0 \quad\quad -p+k+j\text{-}j\text{-}1 \quad 0 \quad\quad 1 \quad\quad\quad 0}{\bullet \cdots \bullet\!\!-\!\!\times\!\!-\!\!\bullet \cdots \bullet \cdots \bullet} \atop \qquad\qquad\qquad\qquad\qquad\uparrow\; p+j^{\text{th}}\text{ node}$$

The hypercohomology spectral sequence has an E_1 term of the form

	0	0		0	0	0			
$p-1^{\text{st}}$ row	*	*	...	*	0	0			
	0	0		0	0	0			
						0			
						0	0	0	0	0
						0	0	*	* ...	*

$\qquad\qquad\qquad\qquad\uparrow \qquad\qquad\qquad\qquad \uparrow$

$\qquad\qquad\qquad -k^{\text{th}}\text{ column} \qquad p-k^{\text{th}}\text{ column}$

and converges after p derivations. When $k \leq -q$ or $k \geq p$, the first level of this sequence has a single row (and one is in the anti-dominant or dominant situation, respectively). Otherwise, λ_k is singular and non-trivial higher differentials can occur in the spectral sequence.

We shall show in chapter 11 that $H^*(Z, \mathcal{O}(k))$ vanishes in degree p or higher for all k. As in remark 9.2.1, this implies the non-vanishing of some of these higher differentials. For example, when $k = 0$, this gives a pth order non-standard differential operator

$$\epsilon_{[A'B'\dots D']} \nabla^{[A'}_{[A} \nabla^{B'}_{B} \cdots \nabla^{D']}_{D]} : \underset{\bullet}{0} \underset{\bullet}{0} \cdots \underset{\bullet}{0} \underset{\times}{{-}p{+}1} \underset{\bullet}{0} \cdots \underset{\bullet}{0} \underset{\bullet}{0}$$

$$\rightarrow \underset{\bullet}{0} \cdots \underset{\bullet}{0} \underset{\times}{{-}p{-}1} \underset{\bullet}{0} \cdots \underset{\bullet}{1} \cdots \underset{\bullet}{0}$$
$$\underset{2p^{\text{th}} \text{ node } \uparrow}{}$$

whose kernel, $H^{p-1}(Z, \mathcal{O}(-p))$, might be thought of as zero rest mass scalar fields.

Now consider the double fibration at (23). As indicated above, the Penrose transform in this context will be of considerable interest in constructing ladder representations of $\mathrm{SU}(p, q)$ in the next chapter. Again, let $X \subset \mathbf{Gr}_p(\mathbf{C}^{p+q})$ be open; it could be either an affine "big cell" or a Stein open orbit of $\mathrm{SU}(p, q)$ such as the orbit of p-planes on which the restriction of the Hermitian form is definite. To continue the notation introduced earlier, we will denote these orbits by $\mathbf{M}^{\pm}$—of course, if $p \neq q$ only $\mathbf{M}^+$ is defined. Let Z be the subset of $\mathbf{CP}^{p+q-1}$ corresponding to X and $\mathbf{P}^{\pm}$ the two open orbits of $\mathrm{SU}(p, q)$.

It is clear that $W^{\mathbf{q}}_{\mathbf{r}}$ is the Hasse subgraph of $W^{\mathbf{P}}$ whose elements do not contain the simple reflection σ_1. (As elements of the symmetric group on $p + q$ letters, they are the elements of $W^{\mathbf{P}}$ which leave the first letter fixed.) Its complement specifies a complex of homogeneous sheaves on X whose cohomology is $H^*(Z, \mathcal{O}_{\mathbf{r}}(\lambda))$, if λ is dominant. For example, the least degree (>0) in which non-trivial cohomology occurs is

$$H^{p-1}(Z, \mathcal{O}_{\mathbf{r}}(\lambda)) \cong \ker \Gamma(X, \mathcal{O}_{\mathbf{p}}(\sigma_p \sigma_{p-1} \cdots \sigma_1.\lambda))$$
$$\rightarrow \Gamma(X, \mathcal{O}_{\mathbf{p}}(\sigma_p \sigma_{p-1} \cdots \sigma_1 \sigma_{p+1}.\lambda)).$$

In fact cohomology in degrees above $p - 1$ vanishes, also; we shall see this in chapter 11—on $\mathbf{P}^+$ this follows from the fact that $\mathbf{P}^+$ is $(p - 1)$-*complete* [54,134].

On the other hand, $w_0 = \sigma_1 \sigma_2 \dots \sigma_{p+q-1}$ is the longest element of $W^{\mathbf{q}}$; it is the cycle $(123 \dots p+q)$ as an element of the symmetric group on $n + 1$ letters. It follows that conjugating an element w of the Hasse diagram $W^{\mathbf{q}}_{\mathbf{r}}$ by w_0:

$$w \rightarrow w' = w_0^{-1} w w_0$$

has the effect of replacing each simple reflection σ_j in w by σ_{j-1}—the result is the Hasse subdiagram of $W_{\mathbf{g}}$ determined by the fibration

Notice that for such a w', $\sigma_p \ldots \sigma_{p+q-1}w' \in W_{\mathbf{r}}^{\mathbf{q}}$ and these exhaust all elements containing σ_{p+q-1}. It follows that

$$\tau_*^{p-1}\mathcal{O}_{\mathbf{q}}(w\sigma_1 \ldots \sigma_{p+q-1}.\lambda) = \mathcal{O}_{\mathbf{p}}(\sigma_p \ldots \sigma_{p+q-1}w')$$

and that the subgraph of elements of $W_{\mathbf{r}}^{\mathbf{q}}$ containing σ_{p+q+1} is a complex whose j^{th} cohomology is $H^j(Z, \mathcal{O}(\sigma_1 \ldots \sigma_{p+q-1}.\lambda))$. This computes the Penrose transform for all antidominant homogeneous bundles on Z.

Example (9.7.1). Let $\lambda = \overset{k}{\times}\!\!-\!\!\overset{0}{\bullet} \ldots \overset{0}{\bullet}\!\!-\!\!\overset{0}{\bullet}$, $k \geq 0$, so $\mathcal{O}_{\mathbf{q}}(\lambda) = \mathcal{O}(k)$. Then

$$H^{p-1}(Z, \mathcal{O}(k))$$

$$= \ker \Gamma(X, \overset{0}{\bullet}\!\!\overset{0}{\bullet} \ldots \overset{0}{\bullet}\!\!\overset{-k-p-1}{\times}\!\!\overset{k+p}{\bullet} \ldots \overset{0}{\bullet}\!\!\overset{0}{\bullet}) \to \Gamma(X, \overset{0}{\bullet} \ldots \overset{1}{\bullet}\!\!\overset{-k-p-2}{\times}\!\!\overset{k+p-1}{\bullet}\!\!\overset{1}{\bullet} \ldots \overset{0}{\bullet})$$

$$= \ker \Gamma(X, \mathcal{O}_{\underbrace{(AB \ldots C)}_{k+p \text{ indices}}}) \to \Gamma(X, \mathcal{O}_{A'\underbrace{(AB \ldots C[D)E]}_{k+p-1 \text{ indices}}}).$$

This extends through the singular cases $k > -p$. Similarly, if $k \leq -p - q$ (the antidominant cases), then

$$H^{p-1}(Z, \mathcal{O}(k))$$

$$= \ker \Gamma(X, \overset{0}{\bullet}\!\!\overset{0}{\bullet} \ldots \overset{k+p-1}{\bullet}\!\!\overset{-k-p}{\times}\!\!\overset{0}{\bullet} \ldots \overset{0}{\bullet}\!\!\overset{0}{\bullet}) \to \Gamma(X, \overset{0}{\bullet} \ldots \overset{1}{\bullet}\!\!\overset{-k-p-1}{\times}\!\!\overset{k+p-2}{\bullet}\!\!\overset{1}{\bullet} \ldots \overset{0}{\bullet})$$

$$= \ker \Gamma(X, \mathcal{O}_{\underbrace{(A'B' \ldots C'D')}_{-k-p \text{ indices}}}[-1]) \to \Gamma(X, \mathcal{O}_{A\underbrace{(A'B' \ldots C'[D')E']}_{-k-p+1 \text{ indices}}}[-1]).$$

Again, this extends through the singular cases $k < -p - 1$ —we leave the verification of this to the reader [47]. The final possibility is

$$H^{p-1}(Z, \mathcal{O}(-p))$$

$$= \ker \Gamma(X, \overset{0}{\bullet}\!\!\overset{0}{\bullet} \ldots \overset{0}{\bullet}\!\!\overset{-1}{\times}\!\!\overset{0}{\bullet} \ldots \overset{0}{\bullet}\!\!\overset{0}{\bullet}) \to \Gamma(X, \overset{0}{\bullet} \ldots \overset{1}{\bullet}\!\!\overset{0}{\bullet}\!\!\overset{-3}{\times}\!\!\overset{0}{\bullet}\!\!\overset{1}{\bullet} \ldots \overset{0}{\bullet})$$

$$= \ker \Gamma(X, \mathcal{O}[-1]) \to \Gamma(X, \mathcal{O}_{[A'B'][AB]}[-1])$$

$$f \longmapsto \nabla_{A'[A}\nabla_{B]B'}f.$$

9.8   An exceptional example

This example of the Penrose transform is more exotic than is usually considered. It is defined between homogeneous spaces for the exceptional group E_6 and is included here purely to emphasize the simple algorithmic form of the transform as described in abstraction in chapters 7 and 8.

In the notation of chapter 7, we shall take

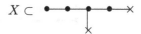

and

As usual, $Z = \eta(\tau^{-1}(X))$ for the correspondence

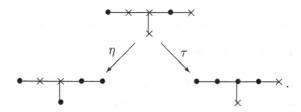

The fibres of η are therefore

$$\times \quad \bullet\!\!-\!\!\times \; = \mathbf{P_1} \times \mathbf{P_2}$$

whilst the fibres of τ are

$$\bullet\!\!-\!\!\times\!\!-\!\!\times\!\!-\!\!\bullet \; = \mathbf{F_{2,3}(C^5)}.$$

Let us take the Penrose transform of the cohomology

$$H^r(Z, \; \overset{1 \;\; -2 \;\; -1 \;\; 1 \;\; 0}{\bullet\!\!-\!\!\times\!\!-\!\!\times\!\!-\!\!\bullet\!\!-\!\!\bullet}\,)$$

assuming, as usual, that the fibres of η have sufficiently simple topology. The relative BGG resolution on $\bullet\!\!-\!\!\times\!\!-\!\!\times\!\!-\!\!\bullet\!\!-\!\!\times$ is

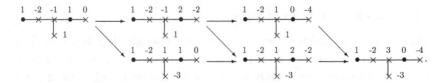

Only first direct images under τ are non-trivial:

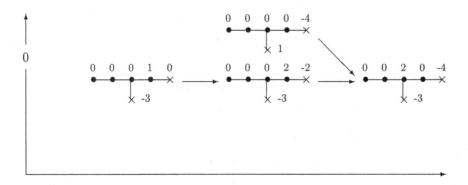

Notice that the operators

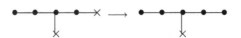

comprise part of a BGG resolution for the mapping

$$\bullet\!\!-\!\!\bullet\!\!-\!\!\bullet\!\!-\!\!\bullet\!\!-\!\!\times \longrightarrow \bullet\!\!-\!\!\bullet\!\!-\!\!\bullet\!\!-\!\!\bullet\!\!-\!\!\bullet$$

so that, for example,

$$H^2(Z, \;\overset{\displaystyle 1\;\;\; -2\;\; -1\;\;\; 1\;\;\;\; 0}{\bullet\!\!-\!\!\times\!\!-\!\!\times\!\!-\!\!\bullet\!\!-\!\!\bullet})$$

$$= \ker : \Gamma(X, \;\overset{0\;\; 0\;\; 0\;\; 1\;\; 0}{\bullet\!\!-\!\!\bullet\!\!-\!\!\bullet\!\!-\!\!\bullet\!\!-\!\!\times}) \to \Gamma(X, \;\overset{0\;\; 0\;\; 0\;\; 2\;\; -2}{\bullet\!\!-\!\!\bullet\!\!-\!\!\bullet\!\!-\!\!\bullet\!\!-\!\!\times})$$

may be interpreted as sections of

$$\overset{0\quad 0\quad 0\quad 1\quad 0}{\bullet\!-\!\bullet\!-\!\bullet\!-\!\bullet\!-\!\bullet}\underset{\times\ -3}{}$$

over the appropriate

open subset of $\bullet\!-\!\bullet\!-\!\bullet\!-\!\bullet\!-\!\bullet$ with $\times$. Similarly, $H^3(Z,\ \overset{1\quad -2\ -1\quad 1\quad 0}{\bullet\!-\!\times\!-\!\times\!-\!\bullet\!-\!\bullet}\underset{\bullet\ 1}{})$ is

identified as sections of

$$\overset{0\quad 0\quad 0\quad 0\quad -4}{\bullet\!-\!\bullet\!-\!\bullet\!-\!\bullet\!-\!\times}\underset{\bullet\ 1}{}$$

under the Penrose transform.

As another example, consider the Penrose transform of the cohomology

$$H^r(Z,\ \overset{0\ -3\ -3\ \ 0\ \ 0}{\bullet\!-\!\times\!-\!\times\!-\!\bullet\!-\!\bullet}\underset{\bullet\ 0}{}).$$

The relative BGG sequence is

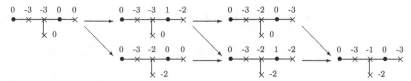

and all, except the first of these, have vanishing direct images under τ. For the first term, we have

$$\tau_*^8\left(\overset{0\ -3\ -3\ \ 0\ \ 0}{\bullet\!-\!\times\!-\!\times\!-\!\bullet\!-\!\times}\underset{\times\ 0}{}\right)=\overset{0\quad 0\quad 0\quad 0\ -3}{\bullet\!-\!\bullet\!-\!\bullet\!-\!\bullet\!-\!\times}\underset{\times\ -6}{}$$

and so

$$H^8(Z,\ \overset{0\ -3\ -3\ \ 0\ \ 0}{\bullet\!-\!\times\!-\!\times\!-\!\bullet\!-\!\bullet}\underset{\bullet\ 0}{})=\Gamma(X,\ \overset{0\quad 0\quad 0\quad 0\ -3}{\bullet\!-\!\times\!-\!\times\!-\!\bullet\!-\!\times}\underset{\times\ -6}{}).$$

A more interesting variation is to consider

$$H^r(Z,\ \overset{0\ -4\ -3\ \ 0\ \ 0}{\bullet\!-\!\times\!-\!\times\!-\!\bullet\!-\!\bullet}\underset{\bullet\ 0}{}).$$

Here, the relative BGG resolution is

and so the spectral sequence of direct images is

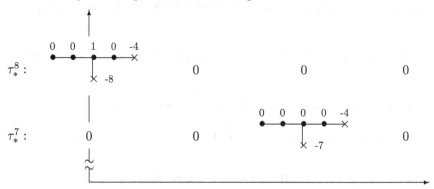

Notice that the fibres of τ are eight-dimensional whence, for X a ball, Z is eight-complete [134] (our conventions follow [144]). Since $H^9(Z, \ \overset{\underset{\displaystyle \bullet\ 0}{}}{\overset{0 \ -4 \ -3 \ 0 \ 0}{\bullet \times \times \bullet \bullet}} \ )$ must therefore vanish, we may deduce the existence of a non-zero (indeed, locally surjective) invariant differential operator

$$D: \ \overset{\underset{\displaystyle \times \ -8}{}}{\overset{0 \ 0 \ 1 \ 0 \ -4}{\bullet \bullet \bullet \bullet \times}} \ \longrightarrow \ \overset{\underset{\displaystyle \times \ -7}{}}{\overset{0 \ 0 \ 0 \ 0 \ -4}{\bullet \bullet \bullet \bullet \times}}$$

whose kernel is $H^8(Z, \ \overset{\underset{\displaystyle \bullet\ 0}{}}{\overset{0 \ -4 \ -3 \ 0 \ 0}{\bullet \times \times \bullet \bullet}} \ )$. This operator D is non-standard and is a good example of an operator whose existence may be deduced through the Penrose transform (see section 8.6).

9.9   The Ward correspondence

The Ward correspondence is a non-linear version of the Penrose transform for the cohomology $H^1(Z, \mathcal{O})$. An element of $H^1(Z, \mathcal{O})$ gives rise to a holomorphic line bundle under the exponential map

$$H^1(Z, \mathcal{O}) \overset{\text{exp}}{\longrightarrow} H^1(Z, \mathcal{O}^*)$$

and the Penrose transform may then be regarded as a construction starting with this line bundle as datum. The Ward correspondence, first developed for the standard case of twistor theory of Minkowski space by Richard Ward [153], takes its datum from a general class of vector bundles rather than just a line bundle.

The theory for a general correspondence as in section 7 is as follows. As observed in section 7.2, a holomorphic bundle E on Z gives rise to a *relative connection*

$$\mathcal{O}(\eta^* E) = \Omega^0_\eta(E) \overset{\nabla_\eta}{\to} \Omega^1_\eta(E)$$

on the pull-back bundle $\eta^* E$ on Y. Conversely, if the fibres of η are connected and simply connected, then a vector bundle on Y with flat relative connection defines a bundle down on Z by means of the covariant constant sections along the fibres. Now suppose that the pull-back $\eta^*(E)$ is trivial on each fibre of τ. Then, the direct image

$$\hat{E} \equiv \tau_* \eta^* E$$

is a vector bundle on X of the same rank as E. Following Manin [113], we shall say that a bundle on Z is *X-trivial* when it has this property. Evidently, this is equivalent to saying that E is trivial on $\eta(\tau^{-1}(x))$ for each $x \in X$. The Ward correspondence arises by considering what happens to the relative connection under direct image by τ. Supposing that $\Gamma(\tau^{-1}(x), \Omega^1_\eta)$ is of constant dimension as x varies, $\tau_* \Omega^1_\eta$ then defines a vector bundle on X and, always, we obtain a first-order differential operator

$$D \equiv \tau_* \nabla_\eta : \hat{E} \to \tau_* \Omega^1_\eta \otimes \hat{E}.$$

This operator satisfies a Leibnitz-type rule induced by the similar property of ∇_η: if ∂ denotes the operator

$$\partial \equiv \tau_* d_\eta : \mathcal{O} \to \tau_* \Omega^1_\eta$$

then

$$D(fs) = f D(s) + \partial f \otimes s.$$

The direct image $\tau_* \Omega^1_\eta$ is closely related to Ω^1_X via a canonical map

$$\Omega^1_X \to \tau_* \Omega^1_\eta.$$

In fact, as pointed out to us by Victor Guillemin, it is straightforward to check that there is a commutative diagram on Y with exact rows and columns

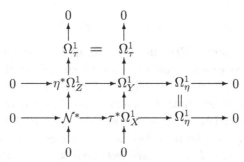

where $\mathcal{N}^*$ is the conormal bundle of Y inside $Z \times X$. The canonical map $\Omega^1_X \to \tau_* \Omega^1_\eta$ is then given by

$$\Omega^1_X = \tau_* \tau^* \Omega^1_X \to \tau_* \Omega^1_\eta$$

provided that τ has connected fibres. This homomorphism of vector bundles relates ∂ to exterior derivative:

$$\mathcal{O} \xrightarrow{\ d\ } \Omega^1$$
$$\partial \searrow \quad \downarrow$$
$$\tau_* \Omega^1_\eta.$$

To proceed further, it is necessary to investigate $\tau_* \Omega^1_\eta$ in greater detail. Fix a point $x \in X$ and write F for the fibre of τ at x. Then F lies inside the submanifold $\eta^{-1}(\eta(F))$ of Y and the normal bundle $\mathcal{V}$ of F inside this submanifold may be regarded as those vectors along F which are tangent to the fibres of η. The derivative of τ induces a mapping of vector bundles

$$\begin{array}{ccc} \mathcal{V} & \xrightarrow{\ \tau'\ } & T \\ \downarrow & & \downarrow \\ F & \xrightarrow{\ \tau\ } & \{x\} \end{array}$$

where T denotes the tangent space to X at x. Notice that $\mathcal{V}$ is not, in general, trivial although it is a subbundle of the trivial bundle $F \times T$. The fibre of the vector bundle $\tau_* \Omega^1_\eta$ at the point x is precisely

$$\Gamma(F, \mathcal{V}^*)$$

and the canonical mapping $\Omega^1 \to \tau_* \Omega^1_\eta$ is given at x by

$$\begin{array}{ccc} T^* & \longrightarrow & \Gamma(F, \mathcal{V}^*) \\ \cup\!\!| & & \cup\!\!| \\ \omega & \longmapsto & \tau^* \omega \end{array}$$

where

$$(\tau^* \omega(y))(v) = \omega(\tau'(v)).$$

Clearly $\ker \tau^* = \mathrm{span}(\tau'(\mathcal{V}))^\circ$ and so τ^* is always non-zero. On the other hand, there is no reason generally that τ^* should be either injective or surjective. Thus, D cannot be related to a genuine connection down on X. Nevertheless, a generalized version of the Ward correspondence goes through as follows. For any $y \in F$, the fibre $\mathcal{V}_y$ of $\mathcal{V}$ at y is injected into T by τ'. But $\mathcal{V}_y$ may be regarded as the tangent vectors to the fibre of η through y. Thus, if D and $\hat{E}$ are restricted to $\tau(\eta^{-1}(z))$ for any $z \in Z$, then we simply recover ∇_η on η^*E. Although D satisfies a Leibnitz-like rule, there is no natural operator

$$\partial : \tau_*\Omega^1_\eta \to \wedge^2\tau_*\Omega^1_\eta$$

and so it is impossible to make sense of a curvature D^2. However, as already observed, D induces a connection on each $\tau(\eta^{-1}(z))$ corresponding to $z \in Z$ and so it makes sense to require that D be flat on each such submanifold. Of course, this is just the same as asking that ∇_η be relatively flat. This completes the circle and we have now proved (with the usual notation and assumptions):

Theorem (9.9.1).   *There is a one-to-one correspondence between:*

- *Differential operators*

$$D : \hat{E} \to \tau_*\Omega^1_\eta \otimes \hat{E}$$

  satisfying
$$D(fs) = fD(s) + \partial f \otimes s$$
  flat on each $\tau(\eta^{-1}(z))$ for all $z \in Z$.

- *X-trivial bundles E of the same rank on Z.*

As an immediate corollary:

Corollary (9.9.2).   *If the natural homomorphism*

$$\Omega^1 \to \tau_*\Omega^1_\eta$$

is an isomorphism, then there is a one-to-one correspondence between:

- *Connections*
$$\nabla : \hat{E} \to \Omega^1 \otimes \hat{E}$$
  flat on each $\tau(\eta^{-1}(z))$ for all $z \in Z$.

- *X-trivial bundles E of the same rank on Z.*

This is the usual and more useful form of the Ward correspondence. To apply it in the case of generalized flag manifolds, it is now necessary specifically to compute $\tau_*\Omega^1_\eta$ for the case of a homogeneous correspondence.

We start by investigating the structure of the cotangent bundle of a general homogeneous manifold G/P. As explained in example 3.2.1, this is the homogeneous bundle arising from the co-Adjoint representation of G on the vector space $(\mathbf{g/p})^*$. For P parabolic, it is possible to study $(\mathbf{g/p})^*$

by identifying it with $\mathbf{u}$, the nilpotent part of $\mathbf{p}$ in a Levi decomposition. The representation is often reducible as in the following typical examples:

$$\Omega^1(\times\!\!\Longrightarrow) = \overset{-2\quad 3}{\times\!\!\Longrightarrow} + \overset{-1\quad 0}{\times\!\!\Longrightarrow}$$

$$\Omega^1(\Longleftarrow\!\!\times) = \overset{1\quad -2}{\Longleftarrow\!\!\times} + \overset{0\quad -1}{\Longleftarrow\!\!\times} + \overset{1\quad -3}{\Longleftarrow\!\!\times}$$

$$\Omega^1(\overset{0\ \ 0\ \ 0\ \ 1\ -2\ \ 1}{\times\!-\!\times\!-\!\bullet\!-\!\bullet\!-\!\times\!-\!\bullet}) = \begin{matrix} \overset{1\ -2\ \ 1\ \ 0\ \ 0\ \ 0}{\times\!-\!\times\!-\!\bullet\!-\!\bullet\!-\!\times\!-\!\bullet} \\[4pt] \overset{-2\ \ 1\ \ 0\ \ 0\ \ 0\ \ 0}{\times\!-\!\times\!-\!\bullet\!-\!\bullet\!-\!\times\!-\!\bullet} \end{matrix} + \begin{matrix} \overset{1\ -1\ \ 0\ \ 0\ -1\ \ 1}{\times\!-\!\times\!-\!\bullet\!-\!\bullet\!-\!\times\!-\!\bullet} \\[4pt] \overset{-1\ -1\ \ 1\ \ 0\ \ 0\ \ 0}{\times\!-\!\times\!-\!\bullet\!-\!\bullet\!-\!\times\!-\!\bullet} \end{matrix} + \overset{-1\ \ 0\ \ 0\ \ 0\ -1\ \ 1}{\times\!-\!\times\!-\!\bullet\!-\!\bullet\!-\!\times\!-\!\bullet}$$

where the notation denotes a composition series with factors as shown. Thus, for example, there is an exact sequence of P-modules:

$$0 \to \overset{-1\quad 0}{\times\!\!\Longrightarrow} \to \Omega^1(\times\!\!\Longrightarrow) \to \overset{-2\quad 3}{\times\!\!\Longrightarrow} \to 0.$$

In general, let $\mathcal{S}_\mathbf{p}$ denote the subset of the simple positive roots $\mathcal{S}$ defining $\mathbf{p}$ as in section 2.2. Recall that $\mathcal{S}_\mathbf{p}$ corresponds to the *uncrossed* nodes on the Dynkin diagram. In the notation of section 2.2, define

$$\pi : \Delta(\mathbf{u}, \mathbf{h}) \to \mathbf{Z}_{\geq 0}^{|\mathcal{S}\backslash\mathcal{S}_\mathbf{p}|}$$

to be taking the coefficients of the roots in $\mathcal{S} \setminus \mathcal{S}_\mathbf{p}$. Let Π denote the image of π. For any $\alpha \in \Delta^+(\mathbf{g}, \mathbf{h})$, define the *height* of α to be the sum of the coefficients of α when written as a linear combination of the simple roots $\mathcal{S}$. For each $\mathbf{n} \in \Pi$, let $\alpha(\mathbf{n}) \in \Delta(\mathbf{u}, \mathbf{h})$ be the element of minimal height such that $\pi(\alpha) = \mathbf{n}$. By considering *root strings* [94], it follows that $\alpha(\mathbf{n})$ is uniquely specified by this requirement and it is easy to see that

$$(\mathbf{g}/\mathbf{p})^* = \sum_{k \geq 1} \left[\bigoplus_{|\mathbf{n}|=k} E(\alpha(\mathbf{n})) \right].$$

Notice that requiring $\alpha(\mathbf{n})$ to be of minimal height implies that $-\alpha(\mathbf{n})$ is dominant for P. This formula is easy to use in practice.

Example (9.9.3).   If $P = $ ⨯⟹, then

$$\Delta(\mathbf{u}, \mathbf{h}) = \{\alpha_1, \alpha_1 + \alpha_2, \alpha_1 + 2\alpha_2, \alpha_1 + 3\alpha_2, 2\alpha_1 + 3\alpha_2\}$$

so

$$\alpha(1) = \alpha_1, \qquad \alpha(2) = 2\alpha_1 + 3\alpha_2.$$

To write these in terms of their weights it is elementary to check that, in general,

$$\alpha = -\sigma_\alpha.0 \text{ for any } \sigma \in \mathcal{S}$$

whence, in this particular case,

$$-\alpha_1 = \overset{-2 \quad 3}{\bullet\!\!=\!\!\!=\!\!\bullet}, \qquad -\alpha_2 = \overset{1 \quad -2}{\bullet\!\!=\!\!\!=\!\!\bullet}$$

so

$$-\alpha(1) = \overset{-2 \quad 3}{\bullet\!\!=\!\!\!=\!\!\bullet}, \qquad -\alpha(2) = -2\alpha_1 - 3\alpha_2 = \overset{-1 \quad 0}{\bullet\!\!=\!\!\!=\!\!\bullet}$$

and we conclude that

$$\Omega^1(\times\!\!\Longrightarrow) = E(\alpha(1)) + E(\alpha(2))$$

$$= \overset{-2 \quad 3}{\times\!\!\Longrightarrow} + \overset{-1 \quad 0}{\times\!\!\Longrightarrow}$$

as above.

Example (9.9.4).   If $P = $ ⨯—⨯—●—●—⨯—●, then

$$\Delta(\mathbf{u}, \mathbf{h}) = \left\{ \begin{array}{l} \alpha_1, \alpha_2, \alpha_5, \alpha_1 + \alpha_2, \alpha_2 + \alpha_3, \alpha_4 + \alpha_5, \alpha_5 + \alpha_6, \\ \alpha_1 + \alpha_2 + \alpha_3, \alpha_2 + \alpha_3 + \alpha_4, \alpha_3 + \alpha_4 + \alpha_5, \alpha_4 + \alpha_5 + \alpha_6, \\ \alpha_1 + \alpha_2 + \alpha_3 + \alpha_4, \alpha_2 + \alpha_3 + \alpha_4 + \alpha_5, \alpha_3 + \alpha_4 + \alpha_5 + \alpha_6, \\ \alpha_1 + \alpha_2 + \alpha_3 + \alpha_4 + \alpha_5, \alpha_2 + \alpha_3 + \alpha_4 + \alpha_5 + \alpha_6, \\ \alpha_1 + \alpha_2 + \alpha_3 + \alpha_4 + \alpha_5 + \alpha_6 \end{array} \right\}$$

so

$$\alpha(1,0,0) = \alpha_1 \qquad\qquad \alpha(0,1,0) = \alpha_2$$
$$\alpha(0,0,1) = \alpha_5 \qquad\qquad \alpha(1,1,0) = \alpha_1 + \alpha_2$$
$$\alpha(0,1,1) = \alpha_2 + \alpha_3 + \alpha_4 + \alpha_5$$
$$\alpha(1,1,1) = \alpha_1 + \alpha_2 + \alpha_3 + \alpha_4 + \alpha_5.$$

We can write these in terms of their weights as in the previous example. For example,

$$-\alpha(0,1,1) = \alpha_2 + \alpha_3 + \alpha_4 + \alpha_5$$

$$= \overset{1 \quad -2 \quad 1 \quad 0 \quad 0 \quad 0}{\bullet\!-\!\bullet\!-\!\bullet\!-\!\bullet\!-\!\bullet\!-\!\bullet} + \overset{0 \quad 1 \quad -2 \quad 1 \quad 0 \quad 0}{\bullet\!-\!\bullet\!-\!\bullet\!-\!\bullet\!-\!\bullet\!-\!\bullet}$$

$$+ \overset{0 \quad 0 \quad 1 \quad -2 \quad 1 \quad 0}{\bullet\!-\!\bullet\!-\!\bullet\!-\!\bullet\!-\!\bullet\!-\!\bullet} + \overset{0 \quad 0 \quad 0 \quad 1 \quad -2 \quad 1}{\bullet\!-\!\bullet\!-\!\bullet\!-\!\bullet\!-\!\bullet\!-\!\bullet}$$

$$+ \overset{1 \quad -1 \quad 0 \quad 0 \quad -1 \quad 1}{\bullet\!-\!\bullet\!-\!\bullet\!-\!\bullet\!-\!\bullet\!-\!\bullet}.$$

These computations give rise to

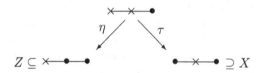

$$\Omega^1(\times\!\!-\!\!\times\!\!-\!\!\bullet\!\!-\!\!\bullet\!\!-\!\!\times\!\!-\!\!\bullet) = \begin{array}{c} 0\ 0\ 0\ 1\ \text{-}2\ 1 \\ \times\!\!-\!\!\times\!\!-\!\!\bullet\!\!-\!\!\bullet\!\!-\!\!\times\!\!-\!\!\bullet \\[4pt] 1\ \text{-}2\ 1\ 0\ 0\ 0 \\ \times\!\!-\!\!\times\!\!-\!\!\bullet\!\!-\!\!\bullet\!\!-\!\!\times\!\!-\!\!\bullet \\[4pt] \text{-}2\ 1\ 0\ 0\ 0\ 0 \\ \times\!\!-\!\!\times\!\!-\!\!\bullet\!\!-\!\!\bullet\!\!-\!\!\times\!\!-\!\!\bullet \end{array} + \begin{array}{c} 1\ \text{-}1\ 0\ 0\ \text{-}1\ 1 \\ \times\!\!-\!\!\times\!\!-\!\!\bullet\!\!-\!\!\bullet\!\!-\!\!\times\!\!-\!\!\bullet \\[4pt] \text{-}1\ \text{-}1\ 1\ 0\ 0\ 0 \\ \times\!\!-\!\!\times\!\!-\!\!\bullet\!\!-\!\!\bullet\!\!-\!\!\times\!\!-\!\!\bullet \end{array} + \begin{array}{c} \text{-}1\ 0\ 0\ 0\ \text{-}1\ 1 \\ \times\!\!-\!\!\times\!\!-\!\!\bullet\!\!-\!\!\bullet\!\!-\!\!\times\!\!-\!\!\bullet \end{array}$$

as above.

Remark (9.9.5).   Notice that these computations are compatible with the BGG resolution of chapter 8. Recall that the first term Δ^1 of this resolution is given by

$$\Delta^1 = \bigoplus_{\alpha \in \mathcal{S} \backslash \mathcal{S}_p} \mathcal{O}(\sigma_\alpha . 0)$$

which is precisely the first term in the composition series above for Ω^1. In other words, there is a surjective homomorphism $\Omega^1 \to \Delta^1$ as already investigated in chapter 8.

It is now possible to compute directly examples of Ω^1_η and their direct images $\tau_* \Omega^1_\eta$ for various cases as follows.

Example (9.9.6).   Consider the double fibration

$$\begin{array}{ccc} & \times\!\!-\!\!\times\!\!-\!\!\bullet & \\ {}^\eta \swarrow & & \searrow {}^\tau \\ Z \subseteq \times\!\!-\!\!\bullet\!\!-\!\!\bullet & & \bullet\!\!-\!\!\times\!\!-\!\!\bullet \supseteq X \end{array}$$

as in section 9.2. The fibre of η is $\times\!\!-\!\!\bullet$ with cotangent bundle

$$\Omega^1(\times\!\!-\!\!\bullet) = \overset{\text{-}2\ \ \ 1}{\times\!\!-\!\!\bullet}.$$

The relative version of previous considerations together with the Bott–Borel–Weil theorem along the fibres of τ now gives

$$\Omega^1_\eta = \overset{1\ \ \text{-}2\ \ 1}{\times\!\!-\!\!\times\!\!-\!\!\bullet} \quad \text{and} \quad \tau_* \Omega^1_\eta = \overset{1\ \ \text{-}2\ \ 1}{\bullet\!\!-\!\!\times\!\!-\!\!\bullet} = \Omega^1(\bullet\!\!-\!\!\times\!\!-\!\!\bullet)$$

so the Ward correspondence is valid in this case. This is the standard example of twistor theory. Notice that the requirement that the connection ∇ generated by the Ward process be flat on the subvarieties $\eta(\tau^{-1}(x))$ (usually referred to as α-*planes* in this example) is re-expressible as a curvature condition. Specifically, the curvature of a relative connection along the fibres of η lies in Ω^2_η and arguing as in the general theory of the Ward correspondence shews that the corresponding connection ∇ down on X is flat along the α-submanifolds if and only if its curvature is annihilated by the natural composition of bundle homomorphisms

$$\Omega^2 \to \wedge^2 \tau_* \Omega^1_\eta \to \tau_* \Omega^2_\eta.$$

It is possible to determine the implications of this condition as follows.

$$\Omega^2_\eta = \wedge^2 (\underset{\times\;\;\;\;\times\;\;\;\;\bullet}{\overset{1\;\;\;\;-2\;\;\;\;1}{\longleftrightarrow}}) = \underset{\times\;\;\;\;\times\;\;\;\;\bullet}{\overset{2\;\;\;\;-3\;\;\;\;0}{\longleftrightarrow}} \;\; \text{so} \;\; \tau_* \Omega^2_\eta = \underset{\bullet\;\;\;\;\times\;\;\;\;\bullet}{\overset{2\;\;\;\;-3\;\;\;\;0}{\longleftrightarrow}}$$

whereas

$$\Omega^2_X = \wedge^2 \Omega^1_X = \wedge^2 (\underset{\bullet\;\;\;\;\times\;\;\;\;\bullet}{\overset{1\;\;\;\;-2\;\;\;\;1}{\longleftrightarrow}}) = \underset{\bullet\;\;\;\;\times\;\;\;\;\bullet}{\overset{2\;\;\;\;-3\;\;\;\;0}{\longleftrightarrow}} \oplus \underset{\bullet\;\;\;\;\times\;\;\;\;\bullet}{\overset{0\;\;\;\;-3\;\;\;\;2}{\longleftrightarrow}} = \Omega^2_+ \oplus \Omega^2_-$$

where the last splitting is into *self-dual* and *anti-self-dual* parts as explained in [127], for example. Thus, a connection on a vector bundle over X is flat on all α-planes if and only if its curvature is anti-self-dual. This is the original form of the Ward correspondence as in [153].

This last argument may, of course, be repeated generally: the Ward connection ∇ will be flat on α-manifolds precisely when its curvature lies in the kernel of

$$\Omega^2 \to \tau_* \Omega^2_\eta.$$

Sometimes this homomorphism is an isomorphism in which case ∇ will be flat and the Ward correspondence of less interest.

Example (9.9.7).   Consider the correspondence

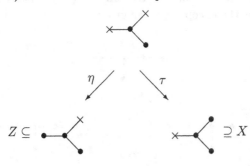

This is the natural twistor correspondence for six dimensional compactified complexified Minkowski space $\mathbf{CS}^6$. In this case we easily compute that

$$\Omega^1_\eta = \overset{-2}{\times}\!\!-\!\!\overset{1}{\bullet}\!\!\diagup\!\!^{\times\,0}_{\bullet\,0} \qquad \text{and} \qquad \Omega^2_\eta = \overset{-3}{\times}\!\!-\!\!\overset{0}{\bullet}\!\!\diagup\!\!^{\times\,1}_{\bullet\,1}$$

so that

$$\tau_*\Omega^1_\eta = \overset{-2}{\times}\!\!-\!\!\overset{1}{\bullet}\!\!\diagup\!\!^{\bullet\,0}_{\bullet\,0} = \Omega^1_X$$

and the Ward correspondence is valid; however,

$$\tau_*\Omega^2_\eta = \overset{-3}{\times}\!\!-\!\!\overset{0}{\bullet}\!\!\diagup\!\!^{\bullet\,1}_{\bullet\,1} = \Omega^2_X$$

so the result is only flat bundles on X. This observation extends to all higher dimensional spheres (and is also the reason why there is no non-trivial analogue of Penrose's non-linear graviton construction [123] in higher dimensions, for conformal manifolds).

It is not always the case that the Ward transform is valid.

Example (9.9.8).　Consider the following correspondence of homogeneous spaces for the exceptional Lie group E_6:

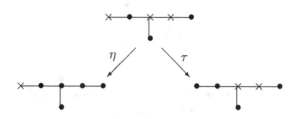

The fibre of η is ●—×—< with cotangent bundle

$$\Omega^1\left(\,\bullet\!-\!\times\!-\!\langle\,\right) = \begin{array}{c} \overset{0\ \ 1\ \ -2}{\bullet\!-\!\times\!-\!\langle}\,{}^1_1 \\[4pt] \oplus \\[4pt] \overset{1\ \ -2\ \ 1}{\bullet\!-\!\times\!-\!\langle}\,{}^0_0 \end{array} \;+\; \overset{1\ \ -1\ \ -1}{\bullet\!-\!\times\!-\!\langle}\,{}^1_1 \;+\; \overset{1\ \ 0\ \ -1}{\bullet\!-\!\times\!-\!\langle}\,{}^0_0 \;+\; \overset{0\ \ -1\ \ 0}{\bullet\!-\!\times\!-\!\langle}\,{}^0_0\,. $$

Thus,

$$\Omega^1_\eta = \begin{array}{c} \overset{0\ \ 1\ \ -2\ \ 1\ \ 0}{\times\!-\!\bullet\!-\!\times\!-\!\times\!-\!\bullet}\,\big|\,1 \\[4pt] \oplus \\[4pt] \overset{0\ \ 0\ \ 1\ \ -2\ \ 1}{\times\!-\!\bullet\!-\!\times\!-\!\times\!-\!\bullet}\,\big|\,0 \end{array} \;+\; \overset{0\ \ 1\ \ -1\ \ -1\ \ 1}{\times\!-\!\bullet\!-\!\times\!-\!\times\!-\!\bullet}\,\big|\,1 \;+\; \overset{1\ \ 0\ \ -1\ \ 0\ \ 1}{\times\!-\!\bullet\!-\!\times\!-\!\times\!-\!\bullet}\,\big|\,0 \;+\; \overset{1\ \ 0\ \ 0\ \ -1\ \ 0}{\times\!-\!\bullet\!-\!\times\!-\!\times\!-\!\bullet}\,\big|\,0 $$

and

$$\tau_*\Omega^1_\eta = \begin{array}{c} \overset{0\ \ 1\ \ -2\ \ 1\ \ 0}{\bullet\!-\!\bullet\!-\!\times\!-\!\times\!-\!\bullet}\,\big|\,1 \\[4pt] \oplus \\[4pt] \overset{0\ \ 0\ \ 1\ \ -2\ \ 1}{\bullet\!-\!\bullet\!-\!\times\!-\!\times\!-\!\bullet}\,\big|\,0 \end{array} \;+\; \overset{0\ \ 1\ \ -1\ \ -1\ \ 1}{\bullet\!-\!\bullet\!-\!\times\!-\!\times\!-\!\bullet}\,\big|\,1 \;+\; \overset{1\ \ 0\ \ -1\ \ 0\ \ 1}{\bullet\!-\!\bullet\!-\!\times\!-\!\times\!-\!\bullet}\,\big|\,0 \;+\; \overset{1\ \ 0\ \ 0\ \ -1\ \ 0}{\bullet\!-\!\bullet\!-\!\times\!-\!\times\!-\!\bullet}\,\big|\,0\,. $$

However, further computations (for which [94] is an invaluable aid) yield

$$\Omega^1_X = \begin{array}{c} \overset{0\ \ 1\ \ -2\ \ 1\ \ 0}{\bullet\!-\!\bullet\!-\!\times\!-\!\times\!-\!\bullet}\,\big|\,1 \\[4pt] \oplus \\[4pt] \overset{0\ \ 0\ \ 1\ \ -2\ \ 1}{\bullet\!-\!\bullet\!-\!\times\!-\!\times\!-\!\bullet}\,\big|\,0 \end{array} \;+\; \overset{0\ \ 1\ \ -1\ \ -1\ \ 1}{\bullet\!-\!\bullet\!-\!\times\!-\!\times\!-\!\bullet}\,\big|\,1 \;+\; \overset{1\ \ 0\ \ -1\ \ 0\ \ 1}{\bullet\!-\!\bullet\!-\!\times\!-\!\times\!-\!\bullet}\,\big|\,0 \;+\; \begin{array}{c} \overset{1\ \ 0\ \ 0\ \ -1\ \ 0}{\bullet\!-\!\bullet\!-\!\times\!-\!\times\!-\!\bullet}\,\big|\,0 \\[4pt] + \\[4pt] \overset{0\ \ 0\ \ -1\ \ 0\ \ 0}{\bullet\!-\!\bullet\!-\!\times\!-\!\times\!-\!\bullet}\,\big|\,1 \end{array} $$

so the natural mapping $\Omega_X^1 \to \tau_* \Omega_\eta^1$ is surjective but not an isomorphism. The usual Ward correspondence therefore fails in this case.

Example (9.9.9).   The Ward correspondence also fails for the double fibration

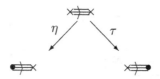

since

$$\tau_* \Omega_\eta^1 = \tau_* (\overset{-2}{\times}\overset{3}{\Longrightarrow}\times) = \overset{-2}{\times}\overset{3}{\Longrightarrow}\bullet$$

whereas

$$\Omega_X^1 = \overset{-2}{\times}\overset{3}{\Longrightarrow}\bullet + \overset{-1}{\times}\overset{0}{\Longrightarrow}\bullet$$

as in example 9.9.3.

It is easy to see from the composition series for Ω_η^1 and Ω_X^1, that $\Omega_X^1 \to \tau_* \Omega_\eta^1$ is *always* surjective for a homogeneous correspondence. Certainly, as in remark 9.9.5, one has an isomorphism

$$\Delta^1 \cong \tau_* \Delta_\eta^1$$

so that the composition series have the same first term. However, the extensions in Ω_η^1 are non-trivial so surjectivity is forced.

To see whether $\Omega_X^1 \to \tau_* \Omega_\eta^1$ is an isomorphism, therefore, it is necessary and sufficient to check the *last* term in the composition series for Ω_X^1 and see whether it also occurs in $\tau_* \Omega_\eta^1$. This is a straightforward task as follows. Consider the dual problem of investigating the first term in the composition series for Θ_X, the tangent bundle to X. The tangent bundle is given by the Adjoint representation of P on

$$\mathbf{g}/\mathbf{p} = \mathbf{u}_- \ (\text{as a vector space})$$

as in example 3.2.1. A lowest weight vector for this is the same as a lowest weight vector for the Adjoint representation of G on $\mathbf{g}$ so the resulting computation is independent of P and we can simply list the results (as in [94]). Thus, the following are the Adjoint representations of the complex semisimple Lie groups:

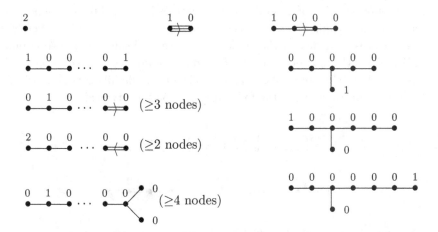

The particular parabolic is hence unimportant in determining the first term in the composition series for $\Theta(G/P)$. For example

$$\Theta(\times\!\!-\!\!\times\!\!\Rightarrow\!\!\bullet\!\!-\!\!\bullet) = \overset{1\quad 0\quad 0\quad 0}{\times\!\!-\!\!\times\!\!\Rightarrow\!\!\bullet\!\!-\!\!\bullet} + \cdots$$

$$\Theta(\bullet\!\!-\!\!\bullet\!\!\Rightarrow\!\!\times\!\!-\!\!\bullet) = \overset{1\quad 0\quad 0\quad 0}{\bullet\!\!-\!\!\bullet\!\!\Rightarrow\!\!\times\!\!-\!\!\bullet} + \cdots .$$

The particular parabolic *is* important in taking duals, however. Thus,

$$\Omega^1(\times\!\!-\!\!\times\!\!\Rightarrow\!\!\bullet\!\!-\!\!\bullet) = \overset{\overset{-2\quad 1\quad 0\quad 0}{\times\!\!-\!\!\times\!\!\Rightarrow\!\!\bullet\!\!-\!\!\bullet}}{\underset{\underset{1\quad -2\quad 2\quad 0}{\times\!\!-\!\!\times\!\!\Rightarrow\!\!\bullet\!\!-\!\!\bullet}}{\oplus}} + \cdots + \overset{-1\quad 0\quad 0\quad 0}{\times\!\!-\!\!\times\!\!\Rightarrow\!\!\bullet\!\!-\!\!\bullet}$$

$$\Omega^1(\bullet\!\!-\!\!\bullet\!\!\Rightarrow\!\!\times\!\!-\!\!\bullet) = \overset{0\quad 1\quad -2\quad 1}{\bullet\!\!-\!\!\bullet\!\!\Rightarrow\!\!\times\!\!-\!\!\bullet} + \cdots + \overset{0\quad 1\quad -2\quad 0}{\bullet\!\!-\!\!\bullet\!\!\Rightarrow\!\!\times\!\!-\!\!\bullet} .$$

In general, the dual of a representation of a semisimple Lie group is computed by considering the action of the longest element of the Weyl group. The result of this consideration is that representations are isomorphic to their duals with the following exceptions where the weights on the Dynkin diagram are reflected as shown:

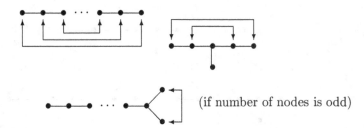

(if number of nodes is odd)

For a representation of a parabolic, this process is applied to the connected
pieces of the Dynkin diagram consisting of uncrossed nodes and the weights
over the crossed nodes are then uniquely determined since W_p is a subgroup
of W_g. Thus, the integers on those crossed nodes which adjoin the uncrossed
ones are altered in a prescribed manner. The precise details are unimportant
for our purposes here but the following examples are typical:

$$\left(\overset{1}{\times}\overset{2}{\text{---}}\overset{3}{\bullet}\text{---}\overset{4}{\bullet}\text{---}\times\right)^* = \overset{-8}{\times}\text{---}\overset{2}{\bullet}\text{---}\overset{3}{\bullet}\text{---}\overset{-14}{\times}$$

$$\left(\overset{1}{\times}\overset{2}{\text{---}}\overset{3}{\bullet}\text{---}\overset{4}{\bullet}\text{---}\overset{5}{\bullet}\text{---}\times\right)^* = \overset{-10}{\times}\text{---}\overset{4}{\bullet}\text{---}\overset{3}{\bullet}\text{---}\overset{2}{\bullet}\text{---}\overset{-23}{\times}.$$

The above information is sufficient to allow a complete determination as
to when the Ward transform is valid. Apart from the obvious condition
that the fibres of η should be connected, the result of this exercise is
unilluminating. Thus, the transform is valid (though trivial) for

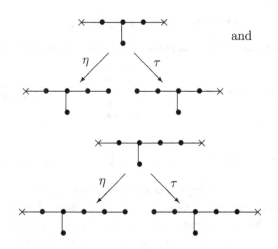

and

but not for

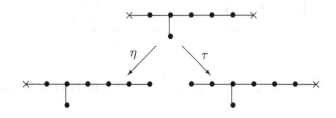

or, indeed, any other homogeneous correspondence for E_8. For $SL(n, \mathbf{C})$, the connectedness of the fibres of η is also sufficient (as previously observed in [50]). For other simple Lie groups, this is not the case as already illustrated in the examples above. We conclude this section with another example for which the transform is valid and gives non-trivial answers.

Example (9.9.10).   Consider the following correspondence for the complex Lie group $SO(10, \mathbf{C})$:

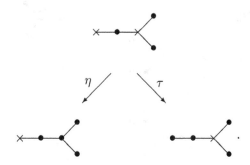

We obtain

$$\Omega^1_\eta = \begin{array}{c} 0 \ \ 1 \ \ -2 \ \ 1 \\ \times\!\!-\!\!\bullet\!\!-\!\!\times\!\!<^{\bullet}_{\bullet} \\ 1 \end{array} + \begin{array}{c} 1 \ \ 0 \ \ -1 \ \ 0 \\ \times\!\!-\!\!\bullet\!\!-\!\!\times\!\!<^{\bullet}_{\bullet} \\ 0 \end{array}$$

and

$$\Omega^1_X = \begin{array}{c} 0 \ \ 1 \ \ -2 \ \ 1 \\ \bullet\!\!-\!\!\bullet\!\!-\!\!\times\!\!<^{\bullet}_{\bullet} \\ 1 \end{array} + \begin{array}{c} 1 \ \ 0 \ \ -1 \ \ 0 \\ \bullet\!\!-\!\!\bullet\!\!-\!\!\times\!\!<^{\bullet}_{\bullet} \\ 0 \end{array}$$

so $\Omega^1_X = \tau_* \Omega^1_\eta$ and the Ward transform is valid. However,

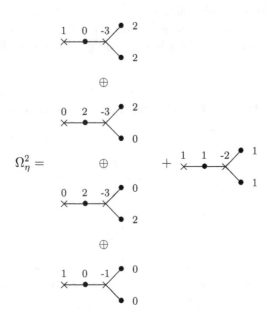

whereas

so the result of the Ward correspondence is connections whose curvature
lies in

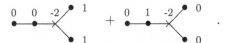

Remark (9.9.11).   If all the fibres $\tau^{-1}(x)$ are isomorphic as in the case
of a homogeneous correspondence, then it makes sense to require that a bundle E on Z belong to the same isomorphism class upon restriction to each
$\eta(\tau^{-1}(x))$. We can develop the Ward correspondence for such X-*uniform*
bundles. The results in certain cases are described in [50,108] and Leiterer
has employed them to study stable bundles on projective spaces [109].

Background coupling

We can also feed the Ward correspondence back into the Penrose transform.
In other words, if E on Z is the Ward transform of a connection ∇ on the
bundle $\hat{E}$ over X, then we can choose any bundle F on Z, for example, a
homogeneous bundle, and inquire as to the interpretation of the cohomology

$$H^r(Z, \mathcal{O}(E \otimes F))$$

in terms of equations on X by means of the Penrose transform. In chapter 7,
the bundle was not assumed homogeneous and so we can apply the general
machinery therein. Recall that if F is homogeneous, then we can use the
Bernstein–Gelfand–Gelfand resolution

$$0 \to \eta^{-1}\mathcal{O}(F) \to \Delta_\eta^\bullet(F)$$

rather than the de Rham resolution in executing the Penrose transform.
For E, however, there is no such alternative and the best we can do is to
use

$$0 \to \eta^{-1}(E \otimes F) \to \Delta_\eta^\bullet(F, E)$$

where $\Delta_\eta^p(F, E) \equiv \Delta_\eta^p(F) \otimes \eta^*(E)$. This is possible because $\eta^*(E)$ is naturally given by transition functions constant along the fibres of η (cf. the case
of the relative de Rham resolution explained at the beginning of section 7.2).

Now recall that the bundle E is X-trivial. In other words, $\eta^*(E)$ is trivial on the fibres of τ. The bundle $\hat{E}$ is simply the direct image $\tau_*\eta^*(E)$. Thus, it is clear that

$$\tau_*^q \Delta_\eta^p(F, E) = \tau_*^q \Delta_\eta^p(F) \otimes \hat{E}$$

and so, in the Penrose transform spectral sequence of theorem 7.3.1, we have

$$E_1^{p,q} = \Gamma(X, \tau_*^q \Delta_\eta^p(F) \otimes \hat{E}) \Longrightarrow H^{p+q}(Z, \mathcal{O}(E \times F))$$

as opposed to

$$E_1^{p,q} = \Gamma(X, \tau_*^q \Delta_\eta^p(F))$$

for the Penrose transform of $H^r(Z, \mathcal{O}(F))$. In other words, it seems as if E is just carried along as a passenger. However, the importance of the Ward correspondence is that the bundle $\hat{E}$ comes equipped with a connection obtained as the direct image of the relative de Rham differential

$$\nabla_\eta : \Omega_\eta^0(E) \to \Omega_\eta^1(E).$$

Bearing in mind that the BGG resolution is also constructed from the relative deRham resolution, it follows that the differential operators which occur in the Penrose transform spectral sequence for the cohomology of $E \otimes F$ are given by precisely the same formulae as for the cohomology of F except that the Levi–Civita connection must be replaced by the tensor product of the Levi–Civita connection with the connection on $\hat{E}$ supplied by the Ward correspondence. In physics, a bundle with connection such as $\hat{E}$ is called a *gauge potential* (strictly, the gauge potential is the connection one-form obtained from a local trivialization (a choice of *gauge*)). A crucial aspect of gauge theory is the realization of the importance of the gauge potential (the connection) as opposed to the *gauge field* (the curvature of this connection). Physicists already know this and mathematicians may find [16] illuminating. If we are given some physically meaningful equations on Minkowski space, such as the massless field equations, then the same equations with the bundle connection replacing the ordinary derivative are known as *minimally coupled* to the background gauge potential. An electromagnetic field provides an example of such a background potential. Indeed, suppose that we consider the standard Penrose transform for Minkowski space as in section 9.2 and suppose that, as on page 103,

$$\omega \in H^1(\mathbf{P}^I, \mathcal{O})$$

corresponds under Penrose transform to $\Phi \in \Gamma(\mathbf{M}^I, \Omega^1)$ such that $d^+\omega = 0$, i.e.,

$$\nabla^A_{(A'}\Phi_{B')A} = 0.$$

Then, whereas the usual massless field equations for positive helicity massless fields read

$$\nabla^{AA'}\varphi_{A'B'\ldots C'} = 0,$$

the background-coupled versions read

$$D^{AA'}\varphi_{A'B'\ldots C'} = 0$$

where $D_a \equiv \nabla_a + \Phi_a$. For negative helicity massless fields, extra care must be taken. For recall that the Penrose transform does not yield fields directly but rather potentials modulo gauge. In the background-coupled case, it is generally no longer true that these are equivalent to fields. For example, in the helicity -1 case (anti-self-dual Maxwell fields) the Penrose transform gives

$$H^1(\mathbf{P}^I, \mathcal{O}) \cong \frac{\ker d^+ \;:\; \Gamma(\mathbf{M}^I, \Omega^1) \to \Gamma(\mathbf{M}^I, \Omega^2_+)}{\operatorname{im} d \;:\; \Gamma(\mathbf{M}^I, \mathcal{O}) \to \Gamma(\mathbf{M}^I, \Omega^1)}$$

and so the background-coupled version for the line bundle $\mathcal{L} = \exp 2\pi i\omega$ is

$$H^1(\mathbf{P}^I, \mathcal{L}) \cong \frac{\ker D^+ \;:\; \Gamma(\mathbf{M}^I, \Omega^1(\mathcal{L})) \to \Gamma(\mathbf{M}^I, \Omega^2_+(\mathcal{L}))}{\operatorname{im} D \;:\; \Gamma(\mathbf{M}^I, \mathcal{L}) \to \Gamma(\mathbf{M}^I, \Omega^1(\mathcal{L}))}$$

where D is the connection on the trivial bundle given by using Φ_a as connection 1-form. The de Rham sequence was employed on page 103 to identify potentials modulo gauge as fields; however, the background-coupled de Rham sequence

$$\Omega^0(\mathcal{L}) \xrightarrow{\;D\;} \Omega^1(\mathcal{L}) \nearrow \begin{array}{c} \Omega^2_+(\mathcal{L}) \\ \\ \Omega^2_-(\mathcal{L}) \end{array} \searrow \Omega^3(\mathcal{L}) \longrightarrow \Omega^4(\mathcal{L})$$

is no longer exact or even a complex, the composition

$$D^2 : \mathcal{L} \to \Omega^2(\mathcal{L})$$

being precisely the curvature. Thus, the very existence of a background-coupled *field* rather than a potential modulo gauge entails some algebraic relationship between the fields and the background curvature. For example,

$$\nabla^{AA'} + \Phi^{AA'} \varphi_{AB} = 0 \quad \Rightarrow \quad F^{AB} \varphi_{AB} = 0$$

where

$$F^{AB} \equiv \nabla_{A'}^{(A} \Phi^{B)A'}$$

is the anti-self-dual part of the curvature or background field induced by the potential Φ_a. In general, such algebraic relations are known as *Buchdahl conditions* [127].

CONSTRUCTING UNITARY REPRESENTATIONS

One of the advantages of the twistor description of positive frequency massless fields via cohomology

$$H^1(\mathbf{PT}^+, \mathcal{O}(k)),$$

as in section 9.2, remark 9.2, and [44,119–121,125], is that it is evident that $SU(2,2)$ is represented on this vector space. This is somewhat simpler than the action of the conformal group $C^\uparrow_+(1,3)$ on the corresponding spacetime fields first noticed by Bateman and Cunningham [14,36] in 1910 for Maxwell's equations and by McLennan [114] in 1956 for general massless fields. This representation is *unitarizable*, i.e., it admits an invariant Hermitian inner product usually called the *scalar product* and, although its definition on massless fields is physically well motivated [64] (in 1961, only informally for spin $\geq \frac{3}{2}$) and [96] (in 1977, with full details), it takes some effort to show its conformal invariance [75] (in 1963) under conformal transformations. However, once the definition of this scalar product has been transferred to cohomology (accomplished in [45,67,68,120]), its invariance is clear. These are the so-called *ladder representations* of $SU(2,2)$ and we can expect the construction to generalize greatly, namely to other groups and their representations on appropriate cohomology. This chapter will describe the extent to which this is justified by means of the cases studied so far.

10.1   The discrete series of $SU(1,1)$

The prototype for any construction concerning semisimple Lie groups is the case of $SL(2,\mathbf{R}) \cong SU(1,1)$. The irreducible unitary representations of this group are completely understood (e.g., [138,147]) and, of these, the so-called *discrete series* representations are often regarded as the most interesting [135].

The usual construction of the discrete series is as follows. SU(1,1) acts on the unit disc $\Delta \equiv \{z \in \mathbf{C} \text{ s.t. } |z| < 1\}$ by factional linear transformations:

$$z \mapsto \frac{az+b}{cz+d} \quad \text{for} \quad \begin{pmatrix} a & b \\ c & d \end{pmatrix} \in \mathrm{SU}(1,1),$$

i.e., for $a = \bar{d}$, $b = \bar{c}$, and $|a|^2 - |b|^2 = 1$. For later use, notice that Δ corresponds to the northern hemisphere

$$S^+ \subset S = \mathbf{CP}_1$$

under stereographic projection from the south pole. As a subset of projective space, S^+ consists of those lines in $\mathbf{C}^2$ on which the Hermitian form preserved by SU(1,1) is positive definite. Now for each integer $n \geq 1$, consider the L^2 Hilbert space

$$\mathcal{H}_n \equiv \left\{ g \in \Gamma(\Delta, \mathcal{O}) \text{ s.t. } \int_\Delta (1 - |z|^2)^{n-1} |f(z)|^2 dx dy < \infty \right\}.$$

This is non-empty and provides a unitary representation of SU(1,1) which acts according to

$$\left[\begin{pmatrix} a & b \\ c & d \end{pmatrix}^{-1} f \right](z) = (cz+d)^{-(n+1)} f\left(\frac{az+b}{cz+d} \right).$$

The simplest case of $n = 1$ can be usefully rephrased as follows. Notice that

$$\int_\Delta |f(z)|^2 dx dy = \frac{1}{2} \int_\Delta i f(z) \overline{f(z)} dz \wedge d\bar{z} = \frac{1}{2} \int_\Delta i \omega \wedge \bar{\omega}$$

where $\omega = f(z) dz$. Moreover, if f transforms as above, then ω transforms by

$$\omega = f(z) dz \mapsto (cz+d)^{-2} f\left(\frac{az+b}{cz+d} \right) dz = f(\alpha(z)) d(\alpha(z)) = \alpha^* \omega$$

where $\alpha(z) = \frac{az+b}{cz+d}$. In other words, $\mathcal{H}_1$ consists of the L^2 holomorphic one-forms on Δ. The natural action of SU(1,1) by pull-back is trivially unitary. These representations are the holomorphic discrete series for SU(1,1). Along with their conjugates $\overline{\mathcal{H}_n}$, they exhaust the discrete series representations.

Another representation which morally belongs to this series (but just fails properly to be square-integrable) is the representation on

$$\mathcal{H}_0 \equiv \left\{ f \in \Gamma(\Delta, \mathcal{O}) \text{ s.t. } \lim_{\epsilon \downarrow 0} \int (1 - |z|^2)^{\epsilon - 1} |f(z)|^2 dx dy < \infty \right\}.$$

An alternative description of these representations which fits rather well with the twistor approach is as follows. Consider the vector space

$$\Gamma(\overline{S^+}, \mathcal{O}).$$

We have the exterior derivative $d : \mathcal{O} \to \Omega^1$ and, hence, an SU(1,1)-invariant transformation

$$d : \Gamma(\overline{S^+}, \mathcal{O}) \to \Gamma(\overline{S^+}, \Omega^1)$$

which is surjective and has $\mathbf{C}$ as kernel. The Hermitian form on $\mathbf{C}^2$ as preserved by SU(1,1) gives rise to an involution of $S = \mathbf{P}(\mathbf{C}^2)$ by

$$\ell \longmapsto \ell^\perp \text{ (orthogonal complement)}.$$

This involution is antiholomorphic, interchanges S^+ and S^-, and has S^0 as fixed points.

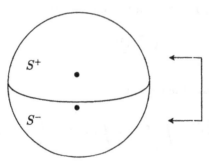

In the usual affine piece $\cong \mathbf{C}$, it is just complex conjugation so the notation of complex conjugation will be maintained by writing $z \longmapsto \bar{z}$. If $f \in \Gamma(\overline{S^+}, \mathcal{O})$, then $f(\bar{z}) \in \Gamma(\overline{S^-}, \overline{\mathcal{O}})$, the antiholomorphic functions on $\overline{S^-}$, whence $\overline{f(\bar{z})}$ is a holomorphic function on $\overline{S^-}$. A similar observation applies to one-forms: if $\omega \in \Gamma(\overline{S^+}, \Omega^1)$, then write $\omega(\bar{z})$ for the pull-back to $\overline{S^-}$. It is an antiholomorphic one-form so $\overline{\omega(\overline{z})}$ is a holomorphic one-form on $\overline{S^-}$. In the usual affine coordinate, this is just

$$g(z)\,dz \longmapsto \overline{g(\overline{z})\,d\overline{z}} = \overline{g(\overline{z})}\,dz.$$

In this way, we obtain conjugate linear isomorphisms

$$\Gamma(\overline{S^+}, \mathcal{O}) \ni f(z) \longmapsto \overline{f(\overline{z})} \in \Gamma(\overline{S^-}, \mathcal{O})$$

$$\Gamma(\overline{S^+}, \Omega^1) \ni \omega \longmapsto \overline{\omega(\overline{z})} \in \Gamma(\overline{S^-}, \Omega^1)$$

which shall simply be written as $f \mapsto \tilde{f}$ and $\omega \mapsto \tilde{\omega}$. These transformations are manifestly SU(1,1)-invariant. We can now define an Hermitian form on $\Gamma(\overline{S^+}, \mathcal{O})$ by

$$\langle f, g \rangle = \oint i f \, \widetilde{dg}$$

where $\oint$ means integrate over S^0, noting that f is defined on $\overline{S^+}$ whilst $\widetilde{dg}$ is defined on $\overline{S^-}$ so that their product is well defined on S^0 as common boundary. Integration by parts gives

$$\langle f, g \rangle = -\oint i \, df \, \tilde{g} = \overline{\oint i g \, \widetilde{df}} = \overline{\langle g, f \rangle}$$

so this form is indeed Hermitian symmetric. Of course, it is not positive definite since it vanishes whenever f or g is constant. In fact, this shows that it is better regarded as an Hermitian form on

$$\frac{\Gamma(\overline{S^+}, \mathcal{O})}{\mathbf{C}} \xrightarrow{\cong} \Gamma(\overline{S^+}, \Omega^1).$$

At this point, Stokes' theorem gives

$$\langle f, g \rangle = \oint i f \, \widetilde{dg} = \oint i f \, \overline{dg} = \int_{S^+} i \, df \wedge \overline{dg}$$

which is just the usual formula for the L^2 inner product on holomorphic one-forms and, hence, this construction recovers the simplest discrete series representation $\mathcal{H}_1$.

   The reason for this somewhat convoluted procedure for deriving what is certainly more transparent in the classical construction is that it generalizes easily to give all the discrete series as follows. Consider the vector space

$$\Gamma(\overline{S^+}, \mathcal{O}(k)) \text{ for } k \geq -1$$

with its SU(1,1) action. The sheaf $\mathcal{O}(k)$ occurs in the very simplest BGG resolution for $S = \mathrm{SL}(2,\mathbf{C})/P = \times$

$$0 \to \overset{k}{\bullet} \to \overset{k}{\times} \overset{\delta}{\to} \overset{-k-2}{\times} \to 0$$

which for $k = 0$ is the de Rham sequence. This differential operator δ is $\eth^{k+1}$, called *edth*, as described in [46]. Thus, we obtain an exact sequence

$$0 \to \odot^k \mathbf{C}^2 \to \Gamma(\overline{S^+}, \mathcal{O}(k)) \overset{\delta}{\to} \Gamma(\overline{S^+}, \mathcal{O}(-k-2)) \to 0$$

equivariant under the action of SU(1,1). There are conjugate linear isomorphisms for all j

$$\Gamma(\overline{S^+}, \mathcal{O}(j)) \ni \nu \mapsto \tilde{\nu} \in \Gamma(\overline{S^-}, \mathcal{O}(j))$$

obtained from complex conjugation in the obvious way generalizing the cases $j = 0$ and $j = -2$ (for Ω^1) used above. Again these isomorphisms are manifestly SU(1,1)-equivariant. The formula for the inner product is now essentially as before:

$$\langle \Phi, \Psi \rangle = \oint i^{k+1} \Phi \widetilde{\delta\Psi},$$

noting that the product $\Phi\widetilde{\delta\Psi}$ is a section of

$$\mathcal{O}(k) \otimes \mathcal{O}(-k-2) = \mathcal{O}(-2) = \Omega^1$$

so that it makes invariant sense to integrate it over S^0. This is Hermitian symmetric by integration by parts and is SU(1,1)-invariant by construction. As before, it vanishes on $\odot^k \mathbf{C}^2$ and so descends to an Hermitian form on $\Gamma(\overline{S^+}, \mathcal{O}(-k-2))$. Thus,

$$\langle F, G \rangle = \oint i^{k+1} \Phi \tilde{G} \text{ where } \delta\Phi = F.$$

Here we should anticipate that it is positive definite. This is readily verified by expanding elements of $\Gamma(\overline{S^+}, \mathcal{O}(-k-2))$ as power series about the north pole. The poles are stabilized by the maximal compact subgroup $K = \mathrm{S}(\mathrm{U}(1) \times \mathrm{U}(1))$ and this power series expansion is precisely an expansion in K-finite vectors of the representation. This expansion also shows that the representations for $k \geq 0$ are indeed the holomorphic discrete series representations $\mathcal{H}_{k+1}$ since the K-types agree. The details are in [47]. Notice that we also obtain a representation when $k = -1$. This is the mock discrete series representation $\mathcal{H}_0$, one which just fails to be in L^2 as indicated earlier. It is obtained with no extra effort.

A rather remarkable consequence of this construction of the discrete series is the following derivation of the *Virasoro algebra*.

Example (10.1.1).   The *Witt* algebra is the algebra of vector fields on the circle. A natural complexification of this is to consider holomorphic vector fields on $S^2 \setminus (\{N\} \cup \{S\})$ (where N and S are the north and south poles). Put $U^+ = S^2 \setminus \{S\}$ and $U^- = S^2 \setminus \{N\}$; then the algebra of interest is

$$V = \Gamma(U^+ \cap U^-, \Theta)$$

where $\Theta = \overset{2}{\times}$. The BGG resolution

$$0 \to \overset{2}{\bullet} \to \overset{2}{\times} \overset{\delta}{\to} \overset{-4}{\times} \to 0$$

yields the following exact sequence in the cohomology of $U^+ \cap U^-$:

$$0 \to \mathbf{sl}(2, \mathbf{C}) \to V \overset{\delta}{\to} \Gamma(U^+ \cap U^-, \overset{-4}{\times}) \to \mathbf{sl}(2, \mathbf{C}) \to 0$$

($\overset{-4}{\times}$ is the sheaf of quadratic differentials on S^2). In this case δ is a third-order operator. Put

$$L_n = z^{-n+1} \frac{d}{dz} \qquad \omega_m = z^{m-2} (dz)^2.$$

Then we have

$$[L_n, L_m] = (n - m) L_{n+m}$$

whilst

$$\delta L_n = n(1 - n^2) \omega_{-n}$$

and so, if $\omega(X, Y) = \langle X, \overline{Y} \rangle$, then

$$\omega(L_n, L_m) = -i \oint z^{-n+1} \frac{d}{dz} \rfloor m(1 - m^2) z^{-m-2} (dz)^2$$

$$= 2\pi m(1 - m^2) \delta_{m, -n}.$$

Clearly ω is an antisymmetric form on V; indeed, it represents a cocycle in the Lie algebra cohomology

$$H^2(V, \mathbf{C})$$

which is non-trivial. The corresponding central extension of V is a complexification of the *Virasoro* algebra (a central extension of the Witt algebra). Writing

$$V = \frac{\Gamma(U^-, \overset{2}{\times})}{\mathbf{sl}(2, \mathbf{C})} \oplus \mathbf{sl}(2, \mathbf{C}) \oplus \frac{\Gamma(U^+, \overset{2}{\times})}{\mathbf{sl}(2, \mathbf{C})}$$

realizes V as a direct sum of $\mathbf{sl}(2, \mathbf{C})$ and two discrete series representations. The central extension comes from the duality of these representations.

10.2   Massless field representations

The definition of the scalar product on massless fields as in [64,75] bears a striking resemblance to the construction of the previous section. The procedure for Maxwell's equations is as follows. Suppose the F is a self-dual solution of Maxwell's equations on $M = real$ compactified Minkowski space (see [117])

$$*F = iF \text{ and } dF = 0$$

or, in usual spinor notation (as in section 9.2),

$$\nabla^{AA'} \phi_{A'B'} = 0 \text{ for } F_{ab} = \phi_{A'B'}\epsilon_{AB}.$$

M is diffeomorphic to $S^1 \times S^3$ so, in particular, $H^2(M, \mathbf{C}) = 0$ and we can find a *potential* Φ for F, namely a one-form Φ such that $d\Phi = F$. Now, if G is another self-dual solution of Maxwell's equations form

$$\oint \Phi \wedge \overline{G}$$

where the integral of this three-form $\Phi \wedge \overline{G}$ is taken over the S^3 of $M = S^1 \times S^3$. Notice first that $\Phi \wedge \overline{G}$ is closed:

$$d(\Phi \wedge \overline{G}) = (d\Phi) \wedge \overline{G} = F \wedge \overline{G}$$

which vanishes because F is self-dual whereas $\overline{G}$ is anti-self-dual. Thus, the particular S^3 is unimportant other than it represents the generator of $H^3(M, \mathbf{Z}) = \mathbf{Z}$. Observe, secondly, that this integral is also independent of choice of potential Φ, for, if Ψ is another, then $d(\Phi - \Psi) = 0$, so in a neighbourhood of S^3 in M we can find f such that $df = \Phi - \Psi$ and now

$$d(f\overline{G}) = \Phi \wedge \overline{G} - \Psi \wedge \overline{G}$$

whence

$$\oint \Phi \wedge \overline{G} = \oint \Psi \wedge \overline{G}$$

by Stokes' theorem. The formula

$$\langle F, G \rangle = \oint \Phi \wedge \overline{G}$$

or, in spinor notation,

$$\langle \phi_{A'B'}, \psi_{A'B'} \rangle = \oint i\Phi_{A'}^{B}\overline{\psi}_{AB}$$

is the definition of the scalar product and bears an obvious resemblance to the formula proposed for the case of SU(1,1). Again, it is manifestly invariant under SO(2,4), the group of orientation preserving conformal automorphisms of M and, if Γ is a potential for G, then

$$\langle F, G \rangle = \oint \Phi \wedge \overline{d\Gamma} = \oint d\Phi \wedge \overline{\Gamma} = \overline{\oint \Gamma \wedge F} = \overline{\langle G, F \rangle}$$

so it is Hermitian symmetric, too. As it stands, this Hermitian form is not positive definite on the space of all self-dual Maxwell fields on M. These Maxwell fields, however, do not constitute an irreducible representation of SO(2,4) but rather split into two irreducibles

{positive frequency self-dual Maxwell fields on M}
$$\oplus$$
{negative frequency self-dual Maxwell fields on M}

as in [5,128,146]. It then turns out that the scalar product is positive definite on one summand and negative definite on the other (precisely which depending on how we choose to orient the S^3 over which we integrate). This is usually seen by Fourier analysing a massless field in terms of plane waves [5,146].

A similar analysis applies for fields of other helicities. In two-spinor notation for helicity $n/2 \geq 1/2$,

$$\langle \phi_{A'B'...D'}, \psi_{A'B'...D'} \rangle = \oint i^{n-1} \Phi_{A'}^{BC...D} \overline{\psi}_{AB...D}$$

where Φ is a potential for ϕ as explained in section 9.2. Notice that this makes good sense because

$$\Phi_{A'BC...D} \in \Gamma(\mathbf{M}, \overset{1 \quad -2 \quad n-1}{\bullet\!\!-\!\!\times\!\!-\!\!\bullet}) \text{ and } \overline{\psi}_{AB...D} \in \Gamma(\mathbf{M}, \overset{0 \quad -n-1 \quad 1}{\bullet\!\!-\!\!\times\!\!-\!\!\bullet})$$

whence $\Phi_{A'}^{BC...D}\overline{\psi}_{AB...D} \in \Gamma(\mathbf{M}, \overset{1 \quad -4 \quad 1}{\bullet\!\!-\!\!\times\!\!-\!\!\bullet}) = \Gamma(\mathbf{M}, \Omega^3)$. For helicity zero, i.e., for solutions of the wave equation, we have

$$\langle \phi, \psi \rangle = \oint i\overline{\psi}\nabla_a \phi - i\phi\nabla_a\overline{\psi}.$$

Our aim in the next section is to show that this seeming analogy with the discrete series realization for SU(1,1) is more than just an analogy. From the twistor point of view, they will both turn out to be special cases of the same procedure.

10.3   The twistor point of view

Recall (section 9.2) that the Penrose transform interprets holomorphic massless fields of helicity $n/2$ on $\mathbf{M}^+$ as cohomology

$$H^1(\mathbf{PT}^+, \mathcal{O}(-n-2))$$

where $\mathbf{PT}^+$ is the open orbit of $\mathbf{PT}$ under SU(2,2) consisting of those lines where the Hermitian form on $\mathbf{T}$ is positive definite. Thus,

$$\mathbf{PT}^+ = \frac{\mathrm{SU}(2,2)}{\mathrm{S}(\mathrm{U}(1) \times \mathrm{U}(1,2))}.$$

M is the Šilov boundary of $\mathbf{M}^+$ and these fields have *hyperfunction* boundary values [5] on M. In this way the cohomology is precisely the positive frequency hyperfunction massless fields of helicity $n/2$ on M. Similarly, the cohomology

$$H^1(\overline{\mathbf{PT}^+}, \mathcal{O}(-n-2))$$

under the Penrose transform is interpreted as the positive frequency *real analytic* massless fields of helicity $n/2$ on M. Thus, we expect the scalar product on massless fields, as in the previous section, to find interpretation on $H^1(\overline{\mathbf{PT}^+}, \mathcal{O}(-n-2))$. Before Penrose introduced cohomology into twistor theory [124,125,128], the twistor description of massless fields was given in terms of integral formulae [118,119,122]. The description of the scalar product in terms of these integral formulae first appeared in [120, pages 277–278]. The cohomological interpretation of this description was accomplished in [45,67,68] and may be given as follows.

The main ingredient is the *twistor transform*

$$\mathcal{T} : H^1(\overline{\mathbf{PT}^+}, \mathcal{O}(-n-2)) \to H^1(\overline{\mathbf{PT}^{*-}}, \mathcal{O}(n-2))$$

obtained by composing the two Penrose transforms identifying the two sides with the holomorphic massless fields of helicity $-n/2$ on $\overline{\mathbf{M}^+}$. If $n \geq 2$, then the left-hand side under the Penrose transform yields these massless fields directly whereas the right-hand side yields a potential modulo gauge description of these same fields (as in section 9.2). Notice that this follows precisely the nature of the classical scalar product formula from the previous section. According to this same formula, the next thing we should do is to take the complex conjugate. On real Minkowski space M, this turns positive frequency into negative frequency and helicity $n/2$ into helicity $-n/2$. Evidently, this is accomplished on cohomology by the conjugate linear isomorphism

$$\sim \; : H^1(\overline{\mathbf{PT}^+}, \mathcal{O}(-n-2)) \xrightarrow{\cong} H^1(\overline{\mathbf{PT}^{*+}}, \mathcal{O}(-n-2))$$

where

$$\tilde{f}(W_\alpha) = \overline{f(\overline{W}^\alpha)}$$

on the level of homogeneous functions. Here, $\overline{W}^\alpha \in \mathbf{T}$ is characterized by

$$\Phi(X_\alpha, W_\alpha) = X_\alpha \overline{W}^\alpha$$

for Φ the given Hermitian form on $\mathbf{T}^*$. In other words, Φ is regarded as a conjugate linear isomorphism $\mathbf{T} \cong \mathbf{T}^*$. To complete the definition of the scalar product, therefore, what is needed is a pairing

$$H^1(\overline{\mathbf{PT}^{*-}}, \mathcal{O}(-n-2)) \otimes_{\mathbf{C}} H^1(\overline{\mathbf{PT}^{*+}}, \mathcal{O}(n-2)) \to \mathbf{C}.$$

On the common boundary, we can pair two classes to obtain an element of

$$H^2(\mathbf{PT}^{*0}, \mathcal{O}(-4)) = H^2(\mathbf{PT}^{*0}, \Omega^3)$$

and now $\mathbf{PT}^{*0}$ is five-dimensional so we can integrate a representative $(3,2)$-form in this cohomology over it to obtain

$$H^2(\mathbf{PT}^{*0}, \Omega^3) \xrightarrow{\oint} \mathbf{C}.$$

This last map may also be regarded as the connecting homomorphism in the Mayer–Vietoris sequence for $\mathbf{PT}^{*\pm}$:

$$H^2(\mathbf{PT}^{*0}, \Omega^3) \to H^3(\mathbf{PT}, \Omega^3) = \mathbf{C}.$$

In this way, the interpretation of the scalar product on massless fields through the Penrose transform becomes

$$\langle \phi, \psi \rangle = \oint (\mathcal{T}\phi)\tilde{\psi}$$

for $\phi, \psi \in H^1(\overline{\mathbf{PT}^+}, \mathcal{O}(-n-2))$.

At this point, we can investigate the formula directly, discarding its origins on Minkowski space. We must show that it is Hermitian symmetric and positive definite. One way of showing that it is Hermitian symmetric is by expressing the twistor transform in terms of *twistor propagators* as in [45] where the resulting formula has the required property by inspection. In this way the twistor transform is constructed directly without having to use fields on Minkowski space. Just as the Penrose transform is a complex analogue of the Radon transform, so the twistor transform is a complex analogue of the Fourier transform. This is explained in [45] and its application in solving differential equations appears in [48]. The construction of the scalar product by means of the twistor transform may be regarded as parallel to the fact that the Fourier transform preserves L^2. An alternative direct construction of the twistor transform without passing through Minkowski space is given in [141]. To show that the scalar product is positive definite from the twistor point of view is more difficult. This is reasonable since there is also some effort needed in the classical construction to show positivity. A good deal of Penrose's original exposition [120] concerns computations with *twistor diagrams*. This was before the introduction of cohomology as the way to understand twistor functions. The resulting calculus may now be viewed as a method of manipulating K-finite vectors which in [120] are termed *elementary states*. (In fact, these are conceptually distinct but turn out to agree; see [54] and below.) Precisely, we can proceed as follows. Fix a maximal compact subgroup

$$S(U(2) \times U(2)) \cong K \subset SU(2,2).$$

Such a subgroup stabilizes a line L in $\mathbf{PT}^-$ and, indeed, the choice of L is precisely the choice of such a K. The complementary line $L^\perp$ in $\mathbf{PT}^+$ is also fixed. For the analogous case of $SU(1,1)$ acting on the sphere, a maximal compact

$$S(U(1) \times U(1)) \cong K \subset SU(1,1)$$

fixes a pair of complementary points, one in each hemisphere $S^\pm$, which may be taken as north and south poles. For the case of $SU(1,1)$, recall that we could show positivity by expanding homogeneous functions on S^+ as Taylor series. The method for $SU(2,2)$ is precisely the same whereby we should expand in K-finite vectors. Thus, one needs to

1. identify the K-finite vectors

2. justify expansion

3. prove positivity.

To accomplish these steps, first consider the cohomology

$$H^1(\mathbf{PT} - L, \mathcal{O}(-n-2)).$$

This is as discussed in section 9.2. It may be computed directly (without reference to spacetime) by choosing two planes whose intersection is L and using their complements as a Leray cover as in section 9.2. See [43,54] and section 11.4 for the details of this computation. The result is that $H^1(\mathbf{PT} - L, \mathcal{O}(-n-2))$ can be described as a set of Laurent series. Since K stabilizes L, it acts on $H^1(\mathbf{PT} - L, \mathcal{O}(-n-2))$ in a manner which may be specifically identified on these Laurent expansions. The K-finite vectors are precisely the finite Laurent series and these form the cohomology $H^1(\mathbf{PT} - L, \mathcal{O}_{\mathrm{alg}}(-n-2))$. Specifically, as K-modules,

if $n \geq 0$,

$$H^1(\mathbf{PT} - L, \mathcal{O}_{\mathrm{alg}}(-n-2)) = \cdots \oplus \overset{n+2 \quad -n-5 \quad 2}{\bullet\!\!-\!\!\times\!\!-\!\!\bullet} \oplus \overset{n+1 \quad -n-3 \quad 1}{\bullet\!\!-\!\!\times\!\!-\!\!\bullet} \oplus \overset{n \quad -n-1 \quad 0}{\bullet\!\!-\!\!\times\!\!-\!\!\bullet}$$

and if $n \leq 0$,

$$H^1(\mathbf{PT} - L, \mathcal{O}_{\mathrm{alg}}(-n-2)) = \cdots \oplus \overset{2 \quad n-5 \quad -n+2}{\bullet\!\!-\!\!\times\!\!-\!\!\bullet} \oplus \overset{1 \quad n-3 \quad -n+1}{\bullet\!\!-\!\!\times\!\!-\!\!\bullet} \oplus \overset{0 \quad n-1 \quad -n}{\bullet\!\!-\!\!\times\!\!-\!\!\bullet} \tag{24}$$

as explained in example 11.4.13 [54]. In standard twistor terminology, an element of one of these irreducible K-representations is called an *elementary state* [43,120] and the theory of *twistor diagrams* [87,88,120] or [91, chapter 5] was initiated by Penrose specifically in order to calculate with them. Since $\mathbf{PT} - L$ is not a union of orbits for SU(2,2), this group does not act on the cohomology $H^1(\mathbf{PT} - L, \mathcal{O}_{\mathrm{alg}}(-n-2))$. However, since $\mathbf{PT} - L$ is open, each of its points has a neighbourhood on which a neighbourhood of the identity in SU(2,2) does act and so the cohomology group is a module over the Lie algebra $\mathbf{su}(2,2)$. Thus, these K-finite vectors $H^1(\mathbf{PT} - L, \mathcal{O}_{\mathrm{alg}}(-n-2))$ form a Harish Chandra module for $\mathbf{su}(2,2)$. There is more on this and on the determination of K-types in chapter 11. Arguments of pseudoconvexity [54] using theorems of Andreotti and Grauert [1] show that

$$H^1(\mathbf{PT} - L, \mathcal{O}(-n-2)) \to H^1(\mathbf{PT}^+, \mathcal{O}(-n-2))$$

is a dense inclusion where $H^1(\mathbf{PT}^+, \mathcal{O}(-n-2))$ is equipped with the natural Fréchet topology induced, for example, by the topology on Dolbeault representatives of uniform convergence of all derivatives on compact subsets [104,140]. This is the *hyperfunction* globalization of the Harish Chandra module $H^1(\mathbf{PT} - L, \mathcal{O}_{\mathrm{alg}}(-n-2))$ as in [40]. The cohomology $H^1(\overline{\mathbf{PT}}^+, \mathcal{O}(-n-2))$ also has a natural topology (of the strong dual of a Fréchet space [54]) and lies between the previous two:

$H^1(\mathbf{PT} - L, \mathcal{O}_{\mathrm{alg}}(-n-2)) = K\text{-finite vectors}$

$\cap$

$H^1(\mathbf{PT} - L, \mathcal{O}(-n-2))$

$\cap$ (25)

$H^1(\overline{\mathbf{PT}^+}, \mathcal{O}(-n-2))$ = "analytic" globalization

$\cap$

$H^1(\mathbf{PT}^+, \mathcal{O}(-n-2))$ = "hyperfunction" globalization

Once the K-finite vectors have been identified as elementary states, it is a simple matter to compute the scalar product on them as in [45,120]. Indeed, the calculus of twistor diagrams is designed for such purposes. It turns out that different K-types are mutually orthogonal whilst the scalar product is positive definite on each specific K-type. Together with the density proved in [54], this shows that the scalar product is at least non-negative definite. We can refine this final argument slightly [54] to prove positivity. The Hilbert space completion of $H^1(\overline{\mathbf{PT}^+}, \mathcal{O}(-n-2))$ is called the space of *normalizable* states of helicity $n/2$ in the physics literature. The very simple pattern of K-types which occurs for these examples (lying on a ray in the weight lattice of K and all occurring with multiplicity one) has given rise to the name *ladder* representations [96].

The above reasoning is not restricted to the case SU(2,2) but applies generally to the case of SU(p,q) for $p, q \geq 2$ acting on

$$H^{p-1}(\overline{\mathbf{P}^+}, \mathcal{O}(-n-p))$$

where $\mathbf{P}^+$ is the orbit of SU(p,q) on $\mathbf{CP}^{p+q-1}$ consisting of those lines in $\mathbf{C}^{p+q}$ on which the invariant Hermitian form is positive definite. The details of the general case are explained in [47]. When p or q is 1, there is only a slight difference in that the module may not be irreducible. Specifically, it may be that $H^0(\mathbf{P}, \mathcal{O}(-n-1))$ is non-zero: this happens when $n \leq -1$. In this case the construction gives an invariant scalar product on the *reduced* cohomology

$$\overline{H}^0(\overline{\mathbf{P}^+}, \mathcal{O}(-n-1)) \equiv \frac{H^0(\overline{\mathbf{P}^+}, \mathcal{O}(-n-1))}{H^0(\mathbf{P}^+, \mathcal{O}(-n-1))}. \qquad (26)$$

When this modification has been incorporated, it can be seen the construction gives the discrete series representations of SU(1,1) as in section 10.1. Thus, the two analogous constructions of sections 10.1 and 10.2 have become special cases of a procedure which should generalize considerably.

Notice that, using the long exact sequence

$$0 = H^0_*(\mathbf{P}^-, \mathcal{O}(-n-1)) \;\rightarrow\; H^0(\mathbf{P}, \mathcal{O}(-n-1)) \rightarrow H^0(\overline{\mathbf{P}^+}, \mathcal{O}(-n-1))$$
$$\rightarrow H^1_*(\mathbf{P}^-, \mathcal{O}(-n-1)) \rightarrow H^1(\mathbf{P}, \mathcal{O}(-n-1)) = 0$$

we can regard $SU(p,q)$ as acting on $H^p_*(\mathbf{P}^-, \mathcal{O}(-n-1))$. This allows $p = 0$, in which case $\mathbf{P} = \mathbf{P}^-$ and $SU(q)$ acts on $H^0(\mathbf{P}, \mathcal{O}(-n))$ according to the Borel–Weil theorem giving the irreducible unitary representation

$$\overset{n}{\bullet}\!-\!\overset{0}{\bullet}\!-\!\overset{0}{\bullet} \;\cdots\; \overset{0}{\bullet}\!-\!\overset{0}{\bullet} \qquad \text{for } n \geq 0.$$

10.4   The twistor transform

The non-trivial ingredient of the previous section was the twistor transform. For $SU(p,q)$, generally,

$$\overline{H}^{p-1}(\mathbf{P}^+, \mathcal{O}(-n-p)) \overset{\mathcal{T}}{\cong} \overline{H}^{q-1}(\mathbf{P}^{*-}, \mathcal{O}(n-q)).$$

The proof [47] is by using the Penrose transform to identify both sides with the kernel of an appropriate holomorphic differential operator on

$$\mathbf{M}^+ = \frac{SU(p,q)}{S(U(p)\times U(q))} \subseteq \mathbf{Gr}_p(\mathbf{C}^{p+q})$$

as for the case $p = q = 2$. We gave the transform for this in example 9.7.1 (at least from $\mathbf{P}^+$ to $\mathbf{M}^+$—the calculation for $\mathbf{P}^{*-}$ is a mirror image). Note that the differential operator involved is standard for $n \neq 0$ and non-standard for $n = 0$.

It is straightforward to apply the Penrose transform of chapters 7 and 9 in order to investigate generally the extent to which the twistor transform is valid. We can try other groups than $SU(p,q)$ and use coefficients in homogeneous vector bundles rather than just line bundles. The case of $SU(2,2)$ (actually $U(2,2)$ for slightly greater generality) and of a general homogeneous vector bundle is treated in [51] where the *generalized* de Rham resolution on $\mathbf{M}$ should be recognized as a BGG resolution. The result proved there is, in the notation of this book, that there is a natural isomorphism

$$\mathcal{T} : H^1(\mathbf{PT}^+, \overset{p}{\times}\!-\!\overset{q}{\bullet}\!-\!\overset{r}{\bullet}) \rightarrow H^1(\mathbf{PT}^{*-}, \overset{q}{\bullet}\!-\!\overset{r}{\bullet}\!-\!\overset{-p\text{-}q\text{-}r\text{-}4}{\times})$$

provided $p \leq -3 - q - r$, $p = -2 - q$, or $p \geq -1$. Combined with the conjugate linear homomorphism

$$H^1(\mathbf{PT}^+, \overset{p}{\times}\!-\!\overset{q}{\bullet}\!-\!\overset{r}{\bullet}) \rightarrow H^1(\mathbf{PT}^{*-}, \overset{r}{\bullet}\!-\!\overset{q}{\bullet}\!-\!\overset{p}{\times}),$$

this gives rise to a pairing on cohomology since

$$\overset{q}{\bullet}\!\!-\!\!\overset{r}{\bullet}\!\!-\!\!\overset{-p-q-r-4}{\times} = \left(\overset{r}{\bullet}\!\!-\!\!\overset{q}{\bullet}\!\!-\!\!\overset{p}{\times}\right)^{*} \otimes \overset{0}{\bullet}\!\!-\!\!\overset{0}{\bullet}\!\!-\!\!\overset{-4}{\times}$$

giving a natural homomorphism

$$\overset{r}{\bullet}\!\!-\!\!\overset{q}{\bullet}\!\!-\!\!\overset{p}{\times} \otimes \overset{q}{\bullet}\!\!-\!\!\overset{r}{\bullet}\!\!-\!\!\overset{-p-q-r-4}{\times} \longrightarrow \overset{0}{\bullet}\!\!-\!\!\overset{0}{\bullet}\!\!-\!\!\overset{-4}{\times} = \Omega^{3}.$$

The K-finite vectors in the cohomology are easily identified but an investigation as to whether the pairing is positive definite has not yet been carried out.

10.5   Hermitian symmetric spaces

As far as other homogeneous spaces are concerned, it seems that there is a valid isomorphism, or twistor transform, whenever we may choose the target space $X \subset G/P$ to be Hermitian symmetric. That is, if G has a real form G_u so that $G_u/K \subset G/P$ is a noncompact Hermitian symmetric space [83], then we can choose R, R' so that there is a twistor transform from $G/R \to G/R'$. The case $G_u = \mathrm{SU}(p,q)$ above is exactly such a situation. The others, amongst classical groups, are in given in table 10.1.

Where only one R is listed the twistor transform acts from G/R to itself. These are precisely the cases in which the variety G/R embeds projectively into a representation F of G which is *self–dual* and in which the conjugation on G over G_u permutes the G_u orbits in G/R. In the other cases, the spaces G/R and G/R' embed projectively into contragredient G-modules and are sent to each other under conjugation. This is exactly the "right" situation, for it will allow us to construct a scalar product.

The twistor transform for these spaces is given in table 10.2 for line bundles. Here we denote by Z (respectively Z^*) the subspace in G/R (respectively, G/R') corresponding to $X = G_u/K \subset G/P$. We let $\mathcal{O}(j)$ on Z or Z^* be the line bundle specified by setting j over the crossed through node in R or R' and zeros over the remainder. Of course, the twistor transform is valid for *any* Z, Z^* corresponding to a Stein or affine $X \subset G/P$. Apart from the cases **C, DIII**, which the reader is invited to check, the calculations needed to verify this table are in section 9.6.

The twistor transform is more generally valid for vector bundles, in the same cohomology degrees as indicated in the table. The bundles are induced by weights with an extreme property relative to $W^{\mathbf{r}}$. Specifically, the transform acts as follows:

$$\mathcal{T} : H^*(Z, \mathcal{O}_{\mathbf{r}}(\lambda)) \overset{\cong}{\longrightarrow} H^*(Z \text{ or } Z^*, \mathcal{O}_{\mathbf{r}}(w_0.\lambda))$$

Table 10.1. Classical Hermitian symmetric spaces and associated twistor spaces for the twistor transform.

	G_u	K	P	R,R'
A	$SU(p,q)$	$S(U(p) \times U(q))$		
B	$SO(2p+1,2)$	$U(1) \times SO(2p+1)$		
C	$Sp(p,\mathbf{R})$	$U(p)$		
DI	$SO(4p,2)$	$U(1) \times SO(4p)$		
DII	$SO(4p+2,2)$	$U(1) \times SO(4p+2)$		
DIII	$SO^*(2p)$	$U(p)$		

Table 10.2. Twistor transform for Hermitian symmetric spaces. $n \geq 0$

A	$H^{p-1}(Z, \mathcal{O}(-p-n)) \cong H^{q-1}(Z^*, \mathcal{O}(-q+n))$
B	$H^{p(p-1)/2}(Z, \mathcal{O}(-2p+1-n)) \cong H^p(Z, \mathcal{O}(-1+n))$
C	$H^{p-1}(Z, \mathcal{O}(-p-1-n)) \cong H^{n-1}(Z, \mathcal{O}(-p+1+n))$
DI	$H^{2p(2p-1)/2}(Z, \mathcal{O}(-4p+2-n)) \cong H^{2p-1}(Z^*, \mathcal{O}(-2+n))$
DII	$H^{(2p+1)p}(Z, \mathcal{O}(-4p-n)) \cong H^{2p}(Z, \mathcal{O}(-2+n))$
DIII	$H^{p-1}(Z, \mathcal{O}(-p-n)) \cong H^{p-2}(Z^*, \mathcal{O}(-p+2+n))$

where $\lambda + \rho$ or $w_0.\lambda + \rho$ is dominant for $\mathbf{g}$ and $w_0 \in W^{\mathbf{q}}$ is the longest element. This covers most of the cases detailed in [51] and explained above.

From this, when $\mathcal{O}_{\mathbf{q}}(\lambda)$ is a line bundle, we should be able to construct unitary representations in the continuation of the holomorphic discrete series for these groups. For line bundles, we expect to obtain ladder representations in all cases but $\mathbf{C}_n$, with K-types occurring along a ray in weight space, with multiplicity one. Notice a common theme amongst all but $\mathbf{B}_n$, namely that at the limit of the transform ($n = 0$ in the table) the spectral sequence in the Penrose transforms used to prove the twistor transform converges only after several derivations. The differential operators that describe the cohomology are *non-standard*. This generalizes the case of $\square$: $\overset{0}{\bullet}\overset{-1}{\underset{\times}{\rule{1.5em}{0.4pt}}}\overset{0}{\bullet} \rightarrow \overset{0}{\bullet}\overset{-3}{\underset{\times}{\rule{1.5em}{0.4pt}}}\overset{0}{\bullet}$ whose kernel is massless scalar fields for $SU(2,2)$ and should be associated with a limiting behaviour in any L^2 construction.

10.6   Towards discrete series

The full extent to which the twistor transform can be defined and is an isomorphism has not yet been determined but, in principle, is simply an exercise in using the Penrose transform of chapter 9. The construction of an invariant sesquilinear pairing now follows. The investigation of these representations and this form by means of their K-finite vectors is still in an early stage but we feel that it is a promising line and deserves further effort.

Recent work by Ed Dunne and one of the authors (MGE) suggests that the twistor transform is valid in a very general setting for orbits of a real form of G on G/B. The calculations strongly suggest that all the discrete series representations and their limits may be constructed directly on the Dolbeault cohomology of these orbits (to complete the approach of Schmid [133,134]).

MODULE STRUCTURES ON COHOMOLOGY

So far, we have really only used representation theory as a rather powerful tool in the calculation of the Penrose transform on an amenable class of manifolds. But the transform is really best thought of as a functor in representation theory, and this chapter turns attention to an attempt to study it as such. The transform should be thought of as a *derived functor*, closely related to the derived *Zuckerman functors* [151] and this is probably the best way for a representation theorist to view what follows. The underlying motivation is quite simple. The Bott–Borel–Weil theorem realizes the cohomology of irreducible homogeneous sheaves on a flag variety G/P as finite dimensional irreducible G-modules. Evidently, cohomology of such sheaves on open proper subsets of G/P cannot bear a G-module structure (since the subset is not preserved under G). But it does bear a $\mathcal{U}(\mathbf{g})$-module structure and, if the subset is an appropriate union of orbits, a K- or G_u-structure, for K a complex semisimple subgroup and G_u a real form of G. These modules vary according as one works in the algebraic or holomorphic category (a new feature), but the results are closely related; indeed they give the geometric equivalent of K-finite vectors as we shall see. As a consequence of the "rigidity" of the holomorphic and algebraic categories, over open *non-affine* or *non-Stein* subvarieties, these modules may have special properties—e.g., they may be Harish Chandra $(\mathbf{g}, K)$-modules, irreducible, etc. On the other hand, over open *affine* or *Stein* subvarieties of G/P, spaces of sections of homogeneous sheaves may be acted on by invariant differential operators, i.e., by operators which intertwine $\mathcal{U}(\mathbf{g})$-module structures. These spaces admit composition series whose terms are kernels and images of such operators. The Penrose transform is the geometric bridge between these two constructions of modules. It begins with cohomology over an open (non-affine) subvariety of one G/R and identifies this in terms of kernels and cokernels of invariant differential operators on sections of irreducible homogeneous sheaves over a different G/P, intertwining $\mathcal{U}(\mathbf{g})$ and K- or G_u-module structures.

Bridges between viewpoints are always useful; this is certainly true of the Penrose transform. The first side of the bridge, involving cohomology, is immediately related to the *relative cohomology* picture occurring in the work of Kempf [98], Beilinson–Bernstein [15] and Brylinski–Kashiwara [28] associating to closed Schubert varieties irreducible quotients of Verma modules. Transport this across the bridge and discover a construction of homomorphisms of Verma modules (including *non-standard* ones) and detailed structural information on Verma modules. Or, if you are a physicist, walk in the opposite direction to discover that spaces of solutions of zero-rest-mass equations (and the solution spaces of several other physically interesting systems of equations) are irreducible quotients of Verma modules, and identify techniques such as *helicity raising and lowering* [128] as standard representation techniques (in this case, the *translation principle*). Various structures on representations are more transparent in one picture or the other. The existence of unitary representations whose K-finite vectors constitute an irreducible quotient of a Verma module [61] exemplifies this; as in the previous chapter, we can utilize the *twistor* transform, (a double Penrose transform) to construct certain unitary representations.

11.1   Verma modules and differential operators

The basic modules which underlie a representation theoretic understanding of the Penrose transform are *Verma modules*, introduced initially by Harish Chandra in [79] and Verma in [150] and studied since then by very many authors—see [17,39,110] for example. Their particular interest for us is that the differential operators that arise in the Penrose transform are induced by homomorphisms of Verma modules. Also, it is often true that the cohomology groups calculated by the Penrose transform are irreducible quotients of Verma modules.

Verma modules

Let $\mathbf{g}$ be a semisimple complex Lie algebra and let $\mathbf{b}$ be a Borel subalgebra. Any finite dimensional irreducible representation of $\mathbf{b}$ is one dimensional; recall from section 3.1 that if $\mathbf{b} = \mathbf{h} \oplus \mathbf{n}$ is a Levi decomposition of $\mathbf{b}$ then $\mathbf{n}$ acts trivially and $\mathbf{h}$ acts by a weight $\lambda \in \mathbf{h}_{\mathbf{R}}^*$. The corresponding representation has been denoted by $F_{\mathbf{b}}(\lambda)$. A Verma module is a $\mathbf{g}$-module *induced* from such a representation, namely, the module

$$V(\lambda) = \mathcal{U}(\mathbf{g}) \otimes_{\mathcal{U}(\mathbf{b})} F_{\mathbf{b}}(\lambda) = \mathsf{ind}_{\mathbf{b}}^{\mathbf{g}} F_{\mathbf{b}}(\lambda)$$

where $\mathbf{g}$ acts by multiplication on the left. The Poincaré–Birkhoff–Witt theorem implies that if $\mathbf{n}_-$ is a complement to $\mathbf{b}$ in $\mathbf{g}$, then

$$V(\lambda) \cong \mathcal{U}(\mathbf{n}_-) \otimes F_{\mathbf{b}}(\lambda) \tag{27}$$

as a vector space, indeed, as an $\mathbf{h}$-module. $V(\lambda)$ is a direct sum of finite dimensional weight spaces. It is clear from this decomposition that the orbit of any element of $V(\lambda)$ under the action of $\mathcal{U}(\mathbf{b})$ is finite dimensional. This fact is usually summarized by saying that $V(\lambda)$ is $\mathcal{U}(\mathbf{b})$-*finite*. The image $\tilde{v}$ of $1 \otimes v$ in $V(\lambda)$, for any non-zero $v \in F_{\mathbf{b}}(\lambda)$, is clearly a highest weight vector. The highest weight space is evidently one dimensional and spanned by $\tilde{v}$ which is annihilated by $\mathbf{n}$; that is, it is a maximal vector. Finally, acting with $\mathcal{U}(\mathbf{n}_-)$, observe that $\tilde{v}$ *generates* $V(\lambda)$.

Relation to differential operators

Suppose now that λ is $\mathbf{g}$-integral. Represent a section of $\mathcal{O}_{\mathbf{b}}(\lambda)$ in the vicinity of the identity coset by an $F_{\mathbf{b}}(-\lambda)$-valued holomorphic function f on G. If the representation is

$$d\pi \; : \; \mathbf{b} \to \mathrm{End}(F_{\mathbf{b}}(-\lambda))$$

and if $\mathcal{U}(\mathbf{g})$ is realised as left invariant differential operators on G, then for $b \in \mathbf{b}$

$$b[f] = -d\pi(b)f.$$

Any differential operator from $F_{\mathbf{b}}(-\lambda)$-valued holomorphic functions on G to holomorphic functions on G is specified by a formula

$$f \mapsto \langle f^*, u[f] \rangle$$

where f^* is an $F_{\mathbf{b}}(\lambda)$-valued function on G and $u \in \mathcal{U}(\mathbf{g})$. When f is a section of $\mathcal{O}_{\mathbf{b}}(\lambda)$ and $b \in \mathbf{b}$, then

$$\langle f^*, b[f] \rangle = \langle f^*, -d\pi(b)f \rangle$$
$$= \langle d\pi^*(b)f^*, f \rangle.$$

Therefore, regarding these differential operators as elements of $\mathcal{U}(\mathbf{g}) \otimes F_{\mathbf{b}}(\lambda)$, at the identity, the action of $b \otimes f^*$ and $1 \otimes d\pi^*(b)f^*$ is the same. In other words, the stalk at the identity coset of the sheaf $\mathcal{D}(\mathcal{O}_{\mathbf{b}}(\lambda), \mathcal{O}_{G/B})$ of linear differential operators from $\mathcal{O}_{\mathbf{b}}(\lambda)$ to $\mathcal{O}_{G/B}$ is identified with $V(\lambda)$. Indeed, $V(\lambda)$, regarded as a *left* $\mathcal{U}(\mathbf{b})$-module by left multiplication, induces this sheaf.

So a good way to visualize Verma modules is as finite order differential operators taking, for example, a holomorphic spinor or tensor field to a holomorphic function.

Duals of Verma modules and jet sheaves

There are two possible modules which we might consider as duals of a Verma module $V(\lambda)$. We shall denote the first of these, the *algebraic dual*, by $V(\lambda)^*$. This is the vector space of all linear functionals on $V(\lambda)$. $\mathbf{g}$ acts contragrediently:

$$x\omega^*(w) = \omega^*(-xw)$$

for $\omega^* \in V(\lambda)^*, w \in V(\lambda)$ and $x \in \mathbf{g}$. The degree filtration on $\mathcal{U}(\mathbf{n}_-)$ induces a filtration on $V(\lambda)$, using (27):

$$V_r(\lambda) = \mathcal{U}(\mathbf{n}_-)_r \otimes_{\mathbf{C}} F_{\mathbf{b}}(\lambda)$$

(so that $V_r(\lambda)/V_{r-1}(\lambda) \cong \odot^r \mathbf{n}_- \otimes F_{\mathbf{b}}(\lambda)$). $V_r(\lambda)^*$ induces the rth jet bundle $J^r(\mathcal{O}_{\mathbf{b}}(\lambda))$. (Recall that sections of $J^r(\mathcal{F})$ consist roughly of sections of $\mathcal{F}$ and their derivatives up to order r.) $V(\lambda)^*$ is the inverse limit of the duals of these:

$$V(\lambda)^* = \varprojlim V_r(\lambda)^*$$

and it induces the *formal jet bundle* $J^\infty(\mathcal{O}_{\mathbf{b}}(\lambda))$ whose sections are formal power series of sections of $\mathcal{O}_{\mathbf{b}}(\lambda)$.

On the other hand, $\mathcal{O}_{\mathbf{b}}(\lambda)$ is a sheaf of $\mathcal{U}(\mathbf{g})$-modules via the left translation action of G. This induces an action of $\mathbf{g}$ on $J^\infty_{eB}(\mathcal{O}_{\mathbf{b}}(\lambda))$ which coincides with the contragredient action of $\mathbf{g}$ on $V(\lambda)^*$. Define

$$_{\mathbf{n}^r_-} V(\lambda)^* = \{\omega \in V(\lambda)^* | \mathbf{n}^r_- \omega = 0\}$$

and hence a direct limit

$$V(\lambda)^\vee = \varinjlim {}_{\mathbf{n}^r_-} V(\lambda)^* \subset V(\lambda)^*.$$

This distinguishes a subbundle of $J^\infty(\mathcal{O}_{\mathbf{b}}(\lambda))$. Sections of $\mathcal{O}_{\mathbf{b}}(\lambda)$ whose jets at the identity coset lie in this subbundle should be thought of as *polynomials*; they are precisely the *algebraic* sections of $\mathcal{O}_{\mathbf{b}}(\lambda)$ which extend to the "big cell" $\exp(\mathbf{n}_-)eB$.

It is easy to describe $V(\lambda)^\vee$ if we pick a weight basis $\{w_i\}$ of $V(\lambda)$ by successively extending bases of $V(\lambda)_r$. Such a basis might be constructed using a weight basis of $\mathbf{n}_-$ and the Poincaré–Birkhoff–Witt theorem. Let $\{\omega_i\} \subset V(\lambda)^*$ be dual to this basis:

$$\omega_i(w_j) = \delta_{ij}.$$

Then $V(\lambda)^\vee = \text{span} \{\omega_i\}$; any element of $V(\lambda)^*$ is a possibly infinite linear combination of ω_i. In particular, $V(\lambda)^\vee$ is a *dense* $\mathbf{g}$-submodule of $V(\lambda)^*$

(relative to the topology in which $K_r = \ker V(\lambda)^* \to V(\lambda)_r^*$ form a basis of open neighbourhoods of zero).

Another characterization of $V(\lambda)^\vee$ is as the **h**-finite vectors in $V(\lambda)^*$; this follows easily from the fact that the weights of **h** on ω_i occur with finite multiplicity. Equivalently, it is the largest **g**-submodule of $V(\lambda)^*$ on which **h** acts semisimply. $V(\lambda)^\vee$ will be called the **h**-finite dual of $V(\lambda)$, or simply a dual Verma module, where no confusion can arise. It is perhaps the more natural notion of dual, here, since $(V(\lambda)^\vee)^\vee = V(\lambda)$. It is also occasionally useful to adopt the notation

$$V(\lambda)^\vee = \text{pro}_{\mathbf{b}}^{\mathbf{g}} F_{\mathbf{b}}(-\lambda)$$
$$\overset{\text{def}}{=} \text{Hom}_{\mathbf{b}}(\mathcal{U}(\mathbf{g}), F_{\mathbf{b}}(-\lambda))_{\mathbf{h}-\text{finite}}.$$

Remark (11.1.1).   [Evaluation.]   The evaluation of a differential operator in $\mathcal{D}(\mathcal{O}_{\mathbf{b}}(\lambda), \mathcal{O}_{G/B})$ on a section of $\mathcal{O}_{\mathbf{b}}(\lambda)$ may be described in terms of Grothendieck duality [74]. If $\dim_{\mathbf{C}}(G/B) = n$ and if $\mathcal{I} \subset \mathcal{O}_{G/B}$ is the ideal sheaf defining the identity coset eB, then for each integer $k \geq 1$, there is a perfect pairing

$$\Gamma(G/B, \mathcal{O}_{\mathbf{b}}(\lambda)/\mathcal{I}^k) \times \text{Ext}_{G/B}^n(\mathcal{O}_{\mathbf{b}}(\lambda)/\mathcal{I}^k, \Omega^n) \to H^n(G/B, \Omega^n) \cong \mathbf{C}.$$

Now $\Gamma(G/B, \mathcal{O}_{\mathbf{b}}(\lambda)/\mathcal{I}^{k+1})$ is just the fibre at eB of the kth associated jet sheaf:

$$\Gamma(G/B, \mathcal{O}_{\mathbf{b}}(\lambda)/\mathcal{I}^{k+1}) \cong J_{eB}^k(\mathcal{O}_{\mathbf{b}}(\lambda)).$$

So, as $k \to \infty$, this identifies

$$\varprojlim \text{Ext}_{G/B}^n(\mathcal{O}_{\mathbf{b}}(\lambda)/\mathcal{I}^k, \Omega^n) \cong \mathcal{D}(\mathcal{O}_{\mathbf{b}}(\lambda), \mathcal{O}_{G/B})_{eB} \cong V(\lambda)$$

and the pairing is just differentiation. In 11.4 a similar construction will give an abstract account of contour integration. (We will say a little more about Grothendieck duality there.)

Elementary properties of Verma modules

A module over $\mathcal{U}(\mathbf{g})$ which is generated by a single maximal highest weight vector is called *standard cyclic*. Such modules possess certain rather special properties [94]:

1. The homomorphic image of any standard cyclic module of highest weight λ is again standard cyclic of highest weight λ, and $V(\lambda)$ is universal amongst these, the evident homomorphism to any other being a surjection.

2. Each standard cyclic module has a unique proper maximal submodule over $\mathcal{U}(\mathbf{g})$, and hence an irreducible quotient; this quotient depends only on λ and is universal in the sense that the evident homomorphism from any other standard cyclic module of weight λ is a surjection. We will denote this quotient by $L(\lambda)$. If λ is dominant integral, then $L(\lambda)$ is the finite dimensional irreducible $\mathbf{g}$-module of highest weight λ. So Verma modules should be thought of as rather more fundamental even than finite dimensional irreducible representations of $\mathbf{g}$. If λ is anti-dominant, then $V(\lambda) = L(\lambda)$. In the category $\mathcal{O}_{\mathbf{b}}$ of finitely generated $\mathcal{U}(\mathbf{b})$-finite $\mathbf{g}$-modules, the $L(\mu)$ are the only irreducible objects. They, or the Verma modules $V(\lambda)$, constitute a basis for the Grothendieck group of this category.

3. The maximal submodule of a standard cyclic module is always a sum of standard cyclic modules; this sum is *not* direct in general. Nonetheless, this means that such modules admit a composition series by irreducible subquotients of the form $L(\mu)$. It is easy to describe all possible μ. For the centre $\mathcal{Z}(\mathbf{g})$ of $\mathcal{U}(\mathbf{g})$ must act on any standard cyclic module by *scalars*, determined entirely by its action on a generating vector. This set of scalars depends, therefore, only on the highest weight λ; thought of as a homomorphism

$$\xi_\lambda \; : \; \mathcal{Z}(\mathbf{g}) \to \mathbf{C}$$

it is usually referred to as the *infinitesimal* or *central character* of the module [79,151]. $L(\mu)$ can therefore occur as a subquotient of a standard cyclic module of highest weight λ only if $\xi_\lambda = \xi_\mu$. This is only possible, by Harish Chandra's theorem [94] if $\mu = w.\lambda$, for $w \in W\mathbf{g}$, the Weyl group of $\mathbf{g}$. Also it is clear by construction that if $L(\mu)$ is a strict subquotient then $\mu \prec \lambda$. These facts imply that, in particular, the composition series has finite length.

4. The $\mathbf{h}$-finite duals $V(\lambda)^\vee$ have unique irreducible submodules, namely $L(\lambda)^\vee$.

Verma modules for other parabolics

A series of particularly important quotients of Verma modules are the *generalized* Verma modules, constructed as follows. Suppose λ is dominant and integral for a parabolic subalgebra $\mathbf{p}$ of $\mathbf{g}$. Then $F_\mathbf{p}(\lambda)$ is a finite dimensional irreducible $\mathbf{p}$-module, dual to $E_\mathbf{p}(\lambda)$, and we may again induce a $\mathbf{g}$-module by

$$M_{\mathbf{p}}(\lambda) = \mathcal{U}(\mathbf{g}) \otimes_{\mathcal{U}(\mathbf{p})} F_{\mathbf{p}}(\lambda).$$

This is standard cyclic, by letting $v \in F_{\mathbf{p}}(\lambda)$ be a highest weight vector and considering the image of $1 \otimes v$. Indeed, if $\mathcal{S}_{\mathbf{p}} \subset \mathcal{S}$ is a set of simple roots defining $\mathbf{p}$, then it is easy to show that there is an exact sequence of $\mathcal{U}(\mathbf{g})$ modules

$$\bigoplus_{\alpha \in \mathcal{S} \backslash \mathcal{S}_{\mathbf{p}}} V(\sigma_\alpha . \lambda) \to V(\lambda) \to M_{\mathbf{p}}(\lambda) \to 0$$

(dual to the initial part of the relative Bernstein–Gelfand–Gelfand resolution).

Again, when λ is $\mathbf{g}$-integral and on G/P, sheaves of differential operators and jet sheaves of homogeneous sheaves are homogeneous sheaves, induced by $M_{\mathbf{p}}(\lambda)$ and $M_{\mathbf{p}}(\lambda)^*$, respectively. A dual generalized Verma module $M_{\mathbf{p}}(\lambda)^\vee$ is defined, in particular, using any complement $\mathbf{u}_-$ of $\mathbf{p}$ in $\mathbf{g}$ (instead of $\mathbf{n}_-$). If $\mathbf{k}$ is a reductive Levi factor of $\mathbf{p}$, then

$$M_{\mathbf{p}}(\lambda)^\vee = M_{\mathbf{p}}(\lambda)^*_{\mathbf{k}\text{–finite}}.$$

11.2   Invariant differential operators

We can now come to an algebraic understanding of the differential operators that occur in the Penrose transform. The key observation is that the whole transform is manifestly invariant under the action of $\mathbf{g}$ so that the differential operators which occur must be G-invariant. But we can classify invariant differential operators in terms of homomorphisms of Verma modules.

Characterization

Recall that a finite order linear differential operator

$$\mathbf{D} \ : \ \mathcal{O}_{\mathbf{p}}(\lambda) \to \mathcal{O}_{\mathbf{p}}(\mu)$$

can be thought of as a homomorphism of sheaves of $\mathcal{O}_{G/P}$ modules

$$\tilde{\mathbf{D}} \ : \ \mathcal{J}^\infty(\mathcal{O}_{\mathbf{p}}(\lambda)) \to \mathcal{O}_{\mathbf{p}}(\mu)$$

which factors through a finite order jet sheaf. Such an operator is called *invariant* if it commutes with the action of G on sections of $\mathcal{O}_{\mathbf{p}}(\lambda)$ and $\mathcal{O}_{\mathbf{p}}(\mu)$. So an invariant differential operator is determined by its action at the identity coset, where it induces a $\mathbf{p}$-module homomorphism

$M_\mathbf{p}(\lambda)^* \to F_\mathbf{p}(\mu)^*$, using the characterization of differential operators given above. Taking adjoints and applying a version of *Frobenius reciprocity*, namely

$$\text{Hom}_\mathbf{p}(F_\mathbf{p}(\mu), M_\mathbf{p}(\lambda)) \cong \text{Hom}_\mathbf{g}(M_\mathbf{p}(\mu), M_\mathbf{p}(\lambda)) \qquad (28)$$

proves

Theorem (11.2.1).   *Invariant linear differential operators between irreducible homogeneous sheaves on the flag variety G/P are in one-to-one correspondence with* **g**-*module homomorphisms of generalized Verma modules*

$$M_\mathbf{p}(\mu) \to M_\mathbf{p}(\lambda).$$

As we said, the differential operators which occur in the Penrose transform are necessarily invariant. So the transform gives a means of constructing homomorphisms of Verma modules. We shall see in 11.5 below that it constructs a great many such homomorphisms, even the rather difficult (*non-standard*) ones. (See section 8.6 and below for an explanation of non-standard.)

Conformally invariant operators

On the other hand, given a Verma module homomorphism we can construct an invariant differential operator which is of some physical interest. Perhaps the most interesting case is for

when the operators are *conformally invariant*. Theorem 11.2.1 and [19,20] classify *all* conformally invariant operators on conformally flat spacetimes [8,53]. We will return to this case below (11.5).

General results

The question of when homomorphisms exist between $V(\mu)$ and $V(\lambda)$ for integral μ, λ is classical [17,150]. The solution may be summarized as:

Theorem (11.2.2).   *Let λ be a dominant integral weight for* **g**. *Then the following statements are equivalent:*

1. *There exists a non-zero injective homomorphism $V(\mu) \to V(\lambda)$, (unique up to scale).*

2. $L(\mu)$ is a subquotient of $V(\lambda)$.

3. $\mu = \sigma_{\beta_n}\sigma_{\beta_{n-1}}\ldots\sigma_{\beta_1}.\lambda$ where $\{\beta_i\}$ is a sequence of positive roots of $\mathbf{g}$ satisfying

   (a) $\langle\lambda+\rho,\beta_1^\vee\rangle > 0$

   (b) For $p = 2,3,\ldots,n$, $\langle\sigma_{\beta_{p-1}}\sigma_{\beta_{p-2}}\ldots\sigma_{\beta_1}.(\lambda+\rho),\beta_p^\vee\rangle > 0$.

The situation for a general parabolic is much more delicate [110]. Suppose there exists a homomorphism from $V(\mu)$ to $V(\lambda)$. We can check that this covers a homomorphism from $M_{\mathbf{p}}(\mu)$ to $M_{\mathbf{p}}(\lambda)$. From the geometrical point of view, this is obvious. We pull back a section of $\mathcal{O}_{\mathbf{p}}(\lambda)$ under the fibration $\tau : G/B \to G/P$, apply the invariant differential operator corresponding to the homomorphism, and take a direct image of the result to obtain a section of $\mathcal{O}_{\mathbf{p}}(\mu)$. This composition is an invariant differential operator on G/P, corresponding to the desired homomorphism.

The homomorphism constructed in this way is called *standard* [110]. It is quite possible for it to be zero. There may, nonetheless, exist a non-zero *non-standard* homomorphism from $M_{\mathbf{p}}(\mu)$ to $M_{\mathbf{p}}(\lambda)$ which is *not* covered by a homomorphism from $V(\mu)$ to $V(\lambda)$.

Example (11.2.3).   Consider $\times\!\!-\!\!\times\!\!-\!\!\times$. Let ∇ be the relative differential of the fibration $\times\!\!-\!\!\times\!\!-\!\!\times \to \times\!\!-\!\!\bullet\!\!-\!\!\times$, and let ∂ and $\tilde{\partial}$ be the relative differentials on the fibrations $\times\!\!-\!\!\times\!\!-\!\!\times \to \bullet\!\!-\!\!\times\!\!-\!\!\times$ and $\times\!\!-\!\!\times\!\!-\!\!\times \to \times\!\!-\!\!\times\!\!-\!\!\bullet$, respectively. Then $\nabla^2\partial\tilde{\partial} : \overset{0\ \ -1\ \ 0}{\times\!\!-\!\!\times\!\!-\!\!\times} \to \overset{0\ \ -3\ \ 0}{\times\!\!-\!\!\times\!\!-\!\!\times}$ is an invariant operator (corresponding to a homomorphism of Verma modules, as in theorem 11.2.2). Evidently, it pushes down to zero between $\overset{0\ \ -1\ \ 0}{\bullet\!\!-\!\!\times\!\!-\!\!\bullet}$ and $\overset{0\ \ -3\ \ 0}{\bullet\!\!-\!\!\times\!\!-\!\!\bullet}$. But the wave operator (see the subsection on page 107)

$$\square : \overset{0\ \ -1\ \ 0}{\bullet\!\!-\!\!\times\!\!-\!\!\bullet} \to \overset{0\ \ -3\ \ 0}{\bullet\!\!-\!\!\times\!\!-\!\!\bullet}$$

is non-zero and corresponds to a non-standard homomorphism of generalized Verma modules.

The existence of non-standard homomorphisms is limited by the following.

Lemma (11.2.4).   [110] *Suppose that there exists a non-zero homomorphism $M_{\mathbf{p}}(\mu) \to M_{\mathbf{p}}(\lambda)$. Then $L(\mu)$ is a subquotient of $M_{\mathbf{p}}(\lambda)$.*

Remark (11.2.5).   The occurrence of non-standard homomorphisms implies the multiple occurrence of $L(\mu)$ as a factor in the composition series for $V(\lambda)$. This suggests that the Penrose transform should give us a better understanding of such composition series.

Remark (11.2.6).   The Penrose transform makes it clear that non-standard homomorphisms of (generalized) Verma modules can also be deduced from homomorphisms of (ordinary) Verma modules. But to make this deduction, we need to consider resolutions such as the Bernstein–Gelfand–Gelfand resolution and higher terms of spectral sequences. In other words, we need somehow to pass to a derived category.

11.3   The algebraic Penrose transform

So now we have a good algebraic grasp of the differential operators which occur in the Penrose transform. The next question is this: can we make algebraic sense of the transform itself? The answer is *yes*. Precisely, the Penrose transform is a *globalization* of Zuckerman's derived functor construction or cohomological parabolic induction [151].

Categories of $\mathbf{g}$–modules and homogeneous sheaves

The first step is to see where the transform should be defined. Let $\mathbf{r} = \mathbf{l} \oplus \mathbf{u} \subset \mathbf{g}$ be a fixed Levi decomposition of a parabolic subalgebra, and suppose $\mathbf{g} = \mathbf{u}_- \oplus \mathbf{r}$. So $\mathbf{r}^t = \mathbf{l} \oplus \mathbf{u}_-$ is a parabolic opposite to $\mathbf{r}$. We may associate two full Abelian subcategories of the category of $\mathbf{g}$-modules to $\mathbf{r}$. The first is $\mathcal{E}_{\mathbf{r}}$ which consists of finitely generated $\mathbf{g}$-modules that become direct sums of finite dimensional irreducible $\mathbf{l}$-modules, on restriction to $\mathbf{l}$. The second, $\mathcal{O}_{\mathbf{r}}$, is the full subcategory of $\mathcal{E}_{\mathbf{r}}$ whose objects are $\mathcal{U}(\mathbf{r})$-locally finite: any element of such a module has a finite dimensional orbit under $\mathcal{U}(\mathbf{r})$. $\mathcal{O}_{\mathbf{r}^t}$ is also defined. $K(\mathcal{O}_{\mathbf{r}})$, the Grothendieck group of $\mathcal{O}_{\mathbf{r}}$ has two distinguished bases, the generalized Verma modules $M_{\mathbf{r}}(\lambda)$ and their irreducible quotients $L(\lambda)$. Similarly, $M_{\mathbf{r}}(\lambda)^{\vee}$ generate $K(\mathcal{O}_{\mathbf{r}^t})$. The notation $\mathcal{O}$ for "holomorphic" and $\mathcal{E}$ for "smooth" betrays its geometric origin.

Also, let $\mathcal{O}_R$ denote the category whose objects are complexes of holomorphic (or regular) homogeneous vector bundles on G/R whose differentials are $\mathbf{C}$–sheaf morphisms which intertwine the action of G. It is too much to ask that these be $\mathcal{O}$-linear because that excludes invariant differential operators. On the other hand, suppose that d is such a differential and that, for any collection $\{f_1, f_2, \ldots, f_n\}$ of holomorphic functions, acting by multiplication, the operator

$$[[\cdots [[d, f_1], f_2], \ldots], f_n] = 0.$$

Then d is an invariant differential operator of order at most n.

Algebraic equivalent of cohomology

The second step is to find an algebraic equivalent of cohomology. By the discussion above, there is a covariant exact functor

$$J_{\mathbf{r}}^{\infty} : \mathcal{O}_R \to \mathcal{O}_{\mathbf{r}^t}.$$

This takes a homogeneous sheaf $\mathcal{F}$ to the l-finite submodule of $J_{eR}^{\infty}(\mathcal{F})$ (with the $\mathbf{g}$ module structure induced by left translation on G):

$$J_{\mathbf{r}}^{\infty}(\mathcal{F}) = \Gamma_1 J_{eR}^{\infty}(\mathcal{F}).$$

Here, Γ_1 is the functor which assigns to a $\mathbf{g}$ module its subset of l-finite vectors. That this subset is a $\mathbf{g}$-submodule is easy to prove—see lemma 11.3.4 below.

   Identify $J_{eR}^{\infty}(\mathcal{F})$ with the stalk at $e \in G$ of F-valued functions on G satisfying the usual invariance property

$$f(gr) = r^{-1} \cdot f(g).$$

Then the trace on $J_{eR}^{\infty}(\mathcal{F})$ of $\Gamma(G/R, \mathcal{F})$ is a finite dimensional $\mathbf{g}$-submodule. Conversely, if $f \in J_{eR}^{\infty}(\mathcal{F})$ is $\mathcal{U}(\mathbf{g})$-finite and so, in some finite dimensional $\mathbf{g}$-submodule F', then F' is a G-module (compatibly with $\mathbf{g}$), since we may assume that G is simply connected. Then f extends to a global section of $\mathcal{F}$ by the formula

$$f(g) = g^{-1}f(e) \text{ so that } f(gr) = r^{-1}g^{-1}f(e) = r^{-1}f(g).$$

This proves that

$$\Gamma(G/R, \cdot) = \Gamma_{\mathbf{g}} \circ J_{\mathbf{r}}^{\infty}(\cdot)$$

so that $\Gamma_{\mathbf{g}}$ is a functor on $\mathcal{O}_{\mathbf{r}^t}$ which corresponds to the global sections functor on sheaves under $J_{\mathbf{r}}^{\infty}$.

   Then the composition of derived functors and the exactness of $J_{\mathbf{r}}^{\infty}$ imply:

Lemma (11.3.1).   *On the category of homogeneous sheaves on G/R,*

$$H^i(G/R, \cdot) = \Gamma_{\mathbf{g}}^i \circ J_{\mathbf{r}}^{\infty}(\cdot)$$

where $\Gamma_{\mathbf{g}}^i$ are the derived functors of $\Gamma_{\mathbf{g}}$.

The algebraic transform and Zuckerman functors

This lemma provides an algebraic equivalent for the push down side of the Penrose transform. Suppose that $\mathbf{p}$ has a reductive Levi factor $\mathbf{k}$. Then the corresponding complex reductive group K acts transitively on the fibre $G/Q \overset{\tau}{\to} G/P$ and it follows from the lemma (in a relative form) that

$$\Gamma_{\mathbf{k}}^i \; : \; \mathcal{O}_{\mathbf{q}^t} \to \mathcal{O}_{\mathbf{p}^t}$$

corresponds, via J^∞, to the direct image τ_*^i:

$$\Gamma_{\mathbf{k}}^i \circ \mathsf{J}_{\mathbf{q}}^\infty \cong \mathsf{J}_{\mathbf{p}}^\infty \circ \tau_*^i.$$

In other words, $\Gamma_{\mathbf{k}}^i$ supplies an algebraic form of the *push down* side of the transform. The left-hand side of the transform is equivalent to the inclusion $\mathcal{O}_{\mathbf{r}^t} \hookrightarrow \mathcal{O}_{\mathbf{q}^t}$. So the *algebraic Penrose transform* is the derived functor (on derived categories):

$$\boxed{\mathbf{RP}_{\mathbf{p}}^{\mathbf{r}} \; : \; \mathcal{O}_{\mathbf{r}^t} \hookrightarrow \mathcal{O}_{\mathbf{q}^t} \overset{\mathrm{R}\Gamma_{\mathbf{k}}}{\to} \mathcal{O}_{\mathbf{p}^t}}$$

Remark (11.3.2).   [Zuckerman functors.]   The derived functor $\mathbf{R}\Gamma_k$ was introduced originally by Zuckerman [151], in 1977 (and explained in his Institute for Advanced Studies lectures in 1978). It is remarkable that the Penrose transform was conceived quite independently (and with no representation theory in mind) by Penrose at about the same time! Closely allied to Zuckerman's functors is his notion of *parabolic induction* defined by

$$\mathcal{R}_{\mathbf{p}}^{\mathbf{g}i} \overset{\text{def}}{=} \Gamma_{\mathbf{k}}^i \circ \mathrm{pro}_{\mathbf{r}}^{\mathbf{g}} (\cdot \otimes \mathsf{D})$$

where $\mathsf{D} = \det \mathbf{u}$ and

$$\mathrm{pro}_{\mathbf{r}}^{\mathbf{g}} (\cdot) = \Gamma_! \mathrm{Hom}_{\mathcal{U}(\mathbf{r})} (\mathcal{U}(\mathbf{g}), \cdot) = (\mathrm{ind}_{\mathbf{r}}^{\mathbf{g}} \cdot)^\vee.$$

If F induces the homogeneous sheaf $\mathcal{F}$, then

$$\mathsf{J}_{\mathbf{r}}^\infty \mathcal{F} = \mathrm{pro}_{\mathbf{r}}^{\mathbf{g}} F,$$

so parabolic induction is the algebraic Penrose transform, composed with tensoring by D.

The Penrose transform itself can be thought of as a derived functor $\mathbf{RP}_P^R$ from $\mathcal{O}_R$ to $\mathcal{O}_P$; we leave the details of this to the reader. Thus

$$\mathbf{RP}_P^R \; = \; \mathbf{R}\tau_* \circ \eta^{-1} \; : \; \mathbf{D}(\mathcal{O}_R) \longrightarrow \mathbf{D}(\mathcal{O}_P).$$

($\mathbf{D}$ means "derived").

The content of the transform lies in the fact that if

$$X = P^t e P \subset G/P \text{ and } Z = P^t e R \subset G/R$$

are corresponding open subvarieties, then

$$\mathbf{R}\Gamma(Z, \cdot) \;=\; \mathbf{R}\Gamma(X, \mathbf{R}P_R^P(\cdot))$$

It is worth remarking that the Penrose transform is *naturally* a derived functor. η^{-1} is *not* a functor between categories of homogeneous sheaves, though it *is* a functor between derived categories, using, say, a BGG or relative de Rham resolution. Then the algebraic and sheaf theoretic Penrose transforms intertwine J^∞:

$$\mathbf{R}\mathsf{P}_\mathbf{r}^\mathbf{P} \circ \mathsf{J}_\mathbf{r}^\infty \;=\; \mathsf{J}_\mathbf{p}^\infty \circ \mathbf{R}\mathsf{P}_R^P. \tag{29}$$

A moment of pure thought shows that on $X \subset G/P$

$$\mathsf{J}_\mathbf{p}^\infty(\cdot) = \Gamma_\mathbf{k} \circ \Gamma(X, \cdot)$$

so (29) becomes

$$\mathbf{R}\mathsf{P}_\mathbf{p}^\mathbf{r} \circ \mathsf{J}_\mathbf{r}^\infty(\cdot) = \Gamma_\mathbf{k} \circ \mathbf{R}\Gamma(Z, \cdot). \tag{30}$$

Remark (11.3.3).   The elements of the cohomology of the right-hand side of (30) are the K-finite vectors of the P^t-modules $H^i(Z, \cdot)$. We shall shortly see another construction of these, which will identify them with the so-called *elementary states* of twistor theory.

Dolbeault resolutions and $\Gamma_\mathbf{g}^i$

It is worth taking a moment to say something about the construction of right derived functors, $\Gamma_\mathbf{g}^i$ of $\Gamma_\mathbf{g}$, from a differential geometric point of view. This will perhaps make them more accessible to physicists. It is easiest to work in the category $\mathcal{E}_\mathbf{r}$. We must construct an injective resolution of each object and then apply $\Gamma_\mathbf{g}$ to the resolution; the derived functors, applied to the object, are the cohomology of the resulting complex. Remarkably, geometry supplies just such a resolution, at least for $M_\mathbf{r}(\lambda)^\vee$. Firstly, recall the following lemma on projective and injective objects in $\mathcal{E}_\mathbf{r}$:

Lemma (11.3.4).   [60,151] *If V is a direct sum of irreducible $\mathbf{l}$-modules, then*

$$\mathrm{ind}_\mathbf{l}^\mathbf{g} V = \mathcal{U}(\mathbf{g}) \otimes_{\mathcal{U}(\mathbf{l})} V$$

and

$$\mathrm{pro}_\mathbf{l}^\mathbf{g} V = \Gamma_\mathbf{l} \mathrm{Hom}_{\mathcal{U}(\mathbf{l})}(\mathcal{U}(\mathbf{g}), V)$$
$$= (\mathrm{ind}_\mathbf{l}^\mathbf{g} V)^\vee$$

are projective and injective objects in $\mathcal{E}_\mathbf{r}$.

Proof   We shall concentrate on $\mathrm{pro}_{\mathbf{l}}^{\mathbf{g}} V$ and leave $\mathrm{ind}_{\mathbf{l}}^{\mathbf{g}} V$ to the reader. Notice that $\mathrm{Hom}_{\mathcal{U}(\mathbf{l})}(\mathcal{U}(\mathbf{g}), V)$ is a **g**-module, by regarding $\mathcal{U}(\mathbf{g})$ as a *right* **g**-module. Now $\mathrm{pro}_{\mathbf{l}}^{\mathbf{g}} V$ is still a **g**-module. To see this, we must check that if $f \in W$ is $\mathcal{U}(\mathbf{l})$-finite then so is vf, for $v \in \mathbf{g}$. But for $l_1, l_2, \ldots l_n \in \mathbf{l}$, commuting gives, for some $v_i \in \mathbf{g}$,

$$l_1 \ldots l_n v f = (v l_1 \ldots l_n f + v_1 l_1 \ldots l_{n-1} f + \ldots + v_n f) \in \mathbf{g} \cdot \mathcal{U}(\mathbf{l}) f.$$

This last subspace is evidently finite dimensional.

To show that $\mathrm{pro}_{\mathbf{l}}^{\mathbf{g}} V$ is injective, we must show that $\mathrm{Hom}_{\mathbf{g}}(\cdot, \mathrm{pro}_{\mathbf{l}}^{\mathbf{g}} V)$ is exact. Let $\mathrm{res}_{\mathbf{g}}^{\mathbf{l}}$ denote the functor which restricts a **g**-module to an **l**-module. Then evaluation at $1 \in \mathcal{U}(\mathbf{g})$ induces a natural equivalence of functors on $\mathcal{E}_{\mathbf{r}}$ *(Frobenius reciprocity)*:

$$\mathrm{Hom}_{\mathbf{g}}(\cdot, \mathrm{pro}_{\mathbf{l}}^{\mathbf{g}} V) \xrightarrow{\cong} \mathrm{Hom}_{\mathbf{l}}(\mathrm{res}_{\mathbf{g}}^{\mathbf{l}}(\cdot), V)$$

so that $\mathrm{res}_{\mathbf{g}}^{\mathbf{l}}$ is a left adjoint of $\mathrm{pro}_{\mathbf{l}}^{\mathbf{g}}$. But $\mathrm{Hom}_{\mathbf{l}}(\cdot, V)$ is exact since **l** is reductive.   □

Up to now, we have been concerned mostly with the structure of G/R as a complex manifold. As a *real* manifold, it is a homogeneous space G_0/L_0 where G_0 is a real compact form of G and $L_0 = R \cap G_0$. (We may assume that **l**, the Lie algebra in G corresponding to L, is the complexification of L_0.) To this real structure are associated sheaves of *smooth* sections of homogeneous vector bundles: the notation $\mathcal{E}_{\mathbf{r}}(\lambda)$ denotes the sheaf associated to $F_{\mathbf{r}}(-\lambda)$, thought of as a complex representation of L_0. We may now argue as in our characterization of differential operators above to show that the sheaf $\mathcal{D}(\mathcal{E}_{\mathbf{r}}(\lambda), \mathcal{E})$ of linear differential operators from $\mathcal{E}_{\mathbf{r}}(\lambda)$ to $\mathcal{E}$ (the sheaf of complex smooth functions) is induced by $\mathrm{ind}_{\mathbf{l}}^{\mathbf{g}}(F_{\mathbf{r}}(\lambda))$ and, hence, that the infinite jet sheaf of smooth sections of $\mathcal{E}_{\mathbf{r}}(\lambda)$ is induced by $\mathrm{ind}_{\mathbf{l}}^{\mathbf{g}}(F_{\mathbf{r}}(\lambda))^*$. Both of these modules are in $\mathcal{E}_{\mathbf{r}}$; in particular, $J_{\mathbf{r}}^{\infty}$ extends to the category of sheaves of smooth sections of homogeneous vector bundles over G/R, with values in $\mathcal{E}_{\mathbf{r}}$. The complexified cotangent bundle of G_0/L_0 is induced by $(\mathbf{g}/\mathbf{l})^*$, as a homogeneous bundle. It follows that $J_{\mathbf{r}}^{\infty}$, sends the *Dolbeault resolution* [157]

$$0 \to \mathcal{O}_{\mathbf{r}}(\lambda) \to \mathcal{E}_{\mathbf{r}}^{0,0}(\lambda) \xrightarrow{\bar{\partial}} \mathcal{E}_{\mathbf{r}}^{0,1}(\lambda) \xrightarrow{\bar{\partial}} \ldots$$

to an injective resolution of $M_{\mathbf{r}}(\lambda)^{\vee}$, in $\mathcal{E}_{\mathbf{r}}$. Here, $\mathcal{E}_{\mathbf{r}}^{p,q}(\lambda)$ is the sheaf of smooth sections of the bundle $\mathcal{O}_{\mathbf{r}}(-\lambda) \otimes \wedge^{p,q} T^*(G_0/L_0))$. This makes lemma (11.3.1) entirely clear.

Remark (11.3.5).   Notice also that $\mathcal{U}(\mathbf{r})$-finiteness corresponds to holomorphicity.

11.4   *K*-types, local cohomology, and elementary states

In section 11.3 we established the formula

$$\Gamma_{\mathbf{k}} \circ \mathbf{R}\Gamma(Z, \cdot) \;=\; \mathbf{R}\mathrm{P}_{\mathbf{r}}^{\mathbf{p}} \circ \mathrm{J}_{\mathbf{r}}$$

where $Z = P^t e R \subset G/R$ was our twistor space, and we worked in the derived category of homogeneous **C**-sheaves on G/R. The cohomology of this functor evaluated on a sheaf $\mathcal{F}$ is the space of K-finite vectors in the P^t-module $H^*(Z, \mathcal{F})$, so that the algebraic Penrose transform may be thought of as halfway to computing the K-types of these representations. These K-finite vectors are the *elementary states* of twistor theory, originally singled out because as cohomology classes they have simple rational function Čech representatives. In this section, we give a brief account of elementary states, as they were originally defined and then in terms of *local cohomology* (a local form of relative cohomology). Then we show how these ideas may be used to calculate K-types.

As before, K is a reductive Levi factor of P; we shall, in fact, assume that $K = P^t \cap P$ where P^t is a parabolic opposite to P (with respect to our fixed choice of Cartan subalgebra and positive roots). If possible, we let G_u be a real form of G whose maximal compact subgroup is a real form of K. To begin with, we take $G = \mathrm{SL}(4, \mathbf{C})$ and $G_u = \mathrm{SU}(2,2)$.

Elementary states

To describe the original construction of elementary states [43], consider the following particular case:

$$H \;=\; H^1(Z, \overset{\text{-n-2}}{\times}\!\!-\!\!\overset{0}{\bullet}\!\!-\!\!\overset{0}{\bullet})$$

where $Z = \times\!\!-\!\!\bullet\!\!-\!\!\bullet \setminus$ line **L**. The Penrose transform identifies H with solutions of the zero rest mass field equations which blow up along the light cone of the point in $\bullet\!\!-\!\!\times\!\!-\!\!\bullet$ corresponding to **L**. If $[Z^\alpha]$ are homogeneous coordinates on $\times\!\!-\!\!\bullet\!\!-\!\!\bullet$ and $A_\alpha, B_\alpha \in F_{\mathbf{g}}(\overset{1}{\bullet}\!\!-\!\!\overset{0}{\bullet}\!\!-\!\!\overset{0}{\bullet}) = \mathbf{T}_\alpha$, then **L** may be taken to be the intersection of the two planes

$$A_\alpha Z^\alpha = 0 \quad \text{and} \quad B_\alpha Z^\alpha = 0. \tag{31}$$

The complements U_A and U_B of these planes together form a Stein cover of Z and so H may be computed with respect to them using Čech cohomology and Laurent series. An element of H is then represented by a function, homogeneous of degree $-n-2$, on $U_A \cap U_B$. To avoid degeneracy, we require these to have poles on the planes at (31):

$$H = \left\{ \sum_{0 < i,j} a_{ij} (A_\alpha Z^\alpha)^{-i} (B_\alpha Z^\alpha)^{-j} \right\}$$

where $a_{ij} \in \Gamma(\mathbf{L}, \mathcal{O}(i+j-n-2))$ and the series is convergent. Following [43], define H_ℓ to be the subspace of H with $a_{ij} = 0$ if $i+j > \ell$ and so obtain a filtration

$$\ldots H_\ell \hookrightarrow H_{\ell+1} \hookrightarrow \ldots \hookrightarrow H.$$

$H_\infty = \lim_{\rightarrow} H_\ell \subset H$ is the set of elementary states. It differs from H only in that H permits infinite series, in the holomorphic category. (In the topology of section 10.3, H_∞ is dense in H.) In the *algebraic* category, there is no difference.

The above filtration and definition of elementary states depends, a priori, on a choice of planes defining $\mathbf{L}$. Invariantly, however, consider instead the algebra

$$\hat{A} = \bigoplus_j \Gamma(\times\!\!-\!\!\bullet\!\!-\!\!\bullet \, , \, \overset{j}{\times}\!\!-\!\!\overset{0}{\bullet}\!\!-\!\!\overset{0}{\bullet})$$

which acts naturally on

$$\hat{H} = \bigoplus_j H^1(Z, \overset{j}{\times}\!\!-\!\!\overset{0}{\bullet}\!\!-\!\!\overset{0}{\bullet}) \supset H.$$

If $\hat{I} \subset \hat{A}$ is the ideal defining $\mathbf{L}$ (generated by $A_\alpha Z^\alpha, B_\alpha Z^\alpha$), then H_ℓ is the subspace annihilated by $\hat{I}^\ell$.

Remark (11.4.1). Using the realization of generalized flag varieties as projective varieties in section 6.1, it is possible to extend the above invariant construction to cohomology on the complement of any subvariety of a generalized flag variety.

$\mathbf{g}$-*module structure on* H_∞

The explicit description of H_∞ just given makes it clear that it bears an action of $\mathrm{sl}(4, \mathbf{C})$ that generates new elementary states from old. To see this, extend A_α, B_α to a basis $\{A_\alpha, B_\alpha, C_\alpha, D_\alpha\}$ of $\mathbf{T}_\alpha$, and consider the following realization of $\mathrm{gl}(4, \mathbf{C})$ by vector fields:

$$\left\{ \begin{array}{cccc} A\partial_A & A\partial_B & A\partial_C & A\partial_D \\ B\partial_A & B\partial_B & B\partial_C & B\partial_D \\ C\partial_A & C\partial_B & C\partial_C & C\partial_D \\ D\partial_A & D\partial_B & D\partial_C & D\partial_D \end{array} \right\} \quad \text{where } A\partial_B \equiv \frac{A_\alpha Z^\alpha \partial}{\partial (B_\beta Z^\beta)}, \text{ etc.}$$

An elementary state in H is a finite linear combination of terms of the form

$$\frac{(C_\alpha Z^\alpha)^p (D_\alpha Z^\alpha)^q}{(A_\alpha Z^\alpha)^i (B_\alpha Z^\alpha)^j} \text{ with } p+q-i-j = -n-2 \text{ and } i,j > 0 \qquad (32)$$

(any such expression with i or $j = 0$ is cohomologous to zero). All of these may be generated under $\mathcal{U}(\mathbf{sl}(4, \mathbf{C}))$ by repeatedly applying the vector fields above to the following elementary states

$$\frac{(C_\alpha Z^\alpha)^{-n}}{(A_\alpha Z^\alpha)(B_\alpha Z^\alpha)} \quad \text{for } n \leq 0$$

$$\frac{1}{(A_\alpha Z^\alpha)(B_\alpha Z^\alpha)^{n+1}} \quad \text{for } n \geq 0.$$

The vector fields given above correspond to right invariant vector fields on SL(4,**C**). Recalling that the action of any G on sections of a homogeneous vector bundle is by $g : f \rightarrow L_{g^{-1}}^* f$ (to obtain a left action), we readily check that the action of

$$-A\partial_A + B\partial_B , \quad -B\partial_B + C\partial_C \text{ and } -C\partial_C + D\partial_D$$

corresponds to the action of the basis elements of our fixed Cartan subalgebra. Thus the positive root spaces of $\mathbf{sl}(4, \mathbf{C})$ are spanned by the vector fields *below* the diagonal. With respect to this ordering, the generating elementary states are lowest weight vectors, of weights

$$\overset{0 \quad -n+1 \quad n}{\bullet\!\!-\!\!\bullet\!\!-\!\!\bullet} \quad \text{when } n \leq 0$$

or

$$\overset{-n \quad n+1 \quad 0}{\bullet\!\!-\!\!\bullet\!\!-\!\!\bullet} \quad \text{when } n \geq 0.$$

So H_∞ is a lowest weight $\mathbf{sl}(4, \mathbf{C})$-module; it is easy to see that it is irreducible (the action of any vector field being reversed by another) and hence an irreducible submodule of a dual Verma module.

Of course, this result is hardly surprising from the point of view of the Penrose transform, since any invariant differential operator determines a submodule of the stalk of the infinite jet bundle at the identity coset, which is a dual Verma module, in this case associated to the lowest weights given above.

Notice that each $H_\ell/H_{\ell-1}$ is actually a module over $K = S(\mathrm{GL}(2, \mathbf{C}) \times \mathrm{GL}(2, \mathbf{C}))$, the complexification of a maximal compact subgroup of $\mathrm{SU}(2, 2)$. Which finite dimensional irreducible representation it is is easy to compute:

$$H_\ell/H_{\ell-1}$$
$$= \mathrm{span}\left\{ \frac{(C_\alpha Z^\alpha)^p (D_\alpha Z^\alpha)^{\ell-n-p-2}}{(A_\alpha Z^\alpha)^i (B_\alpha Z^\alpha)^{\ell-i}} \mid 1 \le i \le \ell-1, 0 \le p \le \ell-n-2 \right\}$$

for $\ell \ge \max{(2, n+2)}$. From the formulae above,

$$\frac{(C_\alpha Z^\alpha)^{\ell-n-2}}{(A_\alpha Z^\alpha)(B_\alpha Z^\alpha)^{\ell-1}} \in H_\ell/H_{\ell-1},$$

is a lowest weight vector for K so that, in our notation,

$$H_\ell/H_{\ell-1} \cong F_\mathbf{p}(\overset{\ell\text{-}2}{\bullet} \overset{n+3\text{-}2\ell}{\times} \overset{\ell\text{-}n\text{-}2}{\bullet}).$$

As a K-module, H_∞ is the direct sum of these. That these constitute *all* the K-types in H will follow from arguments given below. We will also recompute this result as an example (11.4.13) of the more general machine we are about to present.

Remark (11.4.2).   Note that the "lowest K-type" in the representation is given by a weight *different* from that inducing the initial homogeneous sheaf, since it occurs in the cohomology of this sheaf. This is analogous to the parameter shift between the weight inducing a homogeneous line bundle $(\mathcal{L}_\lambda)$ and the lowest K-type in the calculations given by Schmid in [133]. We shall see below that our calculations and his are essentially dual.

Local cohomology

We now want to find a method for computing "elementary states" or K-finite vectors in the general $H^*(Z, \mathcal{F})$ which we can compute using the Penrose transform. To do this we need to *localize* the invariant definition of elementary states given above.

The evident analogues of $\hat{A}$ and $\hat{I}$ are $\mathcal{O}_{G/R}$ and the sheaf of ideals $\mathcal{I} \subset \mathcal{O}_{G/R}$ defining $V \subset G/R$. Given a sheaf $\mathcal{F}$ of $\mathcal{O}_{G/R}$-modules, the calculation above suggests that we should consider sections of $\mathcal{F}$ annihilated by $\mathcal{I}^\ell$ for some ℓ. Define a subsheaf by

$$\mathcal{I}^\ell \mathcal{F}(U) \ = \ \{f \in \mathcal{F}(U) | \mathcal{I}^\ell f = 0\}$$

and observe that $\mathcal{I}^\ell \mathcal{F} \cong \mathcal{H}om_{\mathcal{O}_{G/R}}(\mathcal{O}_{G/R}/\mathcal{I}^\ell, \mathcal{F})$. The phrase "for some ℓ" indicates that we should consider

$$\varinjlim \mathcal{H}om_{\mathcal{O}_{G/R}}(\mathcal{O}_{G/R}/\mathcal{I}^\ell, \mathcal{F}) \subset \mathcal{F}.$$

Since $\mathcal{I} = \mathcal{O}_{G/R}$ away from V, this sheaf is a subsheaf of the sheaf $\underline{\Gamma}_V(\mathcal{F})$ consisting of local sections of $\mathcal{F}$ with support on V; indeed, in the *algebraic* category, they are isomorphic [76, Theorem 2.8]. Because this case is so important, we will adapt our notation to it and set

$$\underline{\Gamma}_{[V]}(\cdot) = \varinjlim \mathcal{H}om_{\mathcal{O}_{G/R}}(\mathcal{O}_{G/R}/\mathcal{I}^\ell, \cdot)$$

Remark (11.4.3).   The sheaf $\mathcal{O}_{(\ell)} = \mathcal{O}_{G/R}/\mathcal{I}^{\ell+1}$ is the structure sheaf of the ℓth *formal neighbourhood* of V in G/R.

If $\mathcal{F}$ is locally free, it is not going to have any sections supported on V. To remedy this situation, we consider as always the derived functor $\mathbf{R}\underline{\Gamma}_{[V]}$ whose cohomology on $\mathcal{F}$ is

$$\mathcal{H}^i_{[V]}(\mathcal{F}) = \varinjlim \mathcal{E}xt^i_{\mathcal{O}_{G/R}}(\mathcal{O}_{G/R}/\mathcal{I}^\ell, \mathcal{F})$$

(recalling that direct limits commute with homological algebra). We should then regard this functor, the *local cohomology* of $\mathcal{F}$ with respect to V, as the local equivalent of the construction of elementary states.

Remark (11.4.4).   [A local differential structure]   Elementary states have a **g**-module structure, as we have seen. This will arise from a local differential structure supplied abstractly by the following observation. Let $\mathcal{D}(\mathcal{F})$ be the sheaf of linear differential operators on $\mathcal{F}$. When $\mathcal{F}$ is locally free, this is defined in the usual way by choosing local trivializations. The Leibnitz rule in $\mathcal{D}(\mathcal{F})$ implies that if $D \in \mathcal{D}(\mathcal{F})$ is of order $\leq i$, then

$$\mathcal{I}^{i+\ell} D\mathcal{F} \subset {}_D\mathcal{I}^\ell \mathcal{F} \tag{33}$$

so that $\Gamma_{[V]}(\mathcal{F})$ has a $\mathcal{D}(\mathcal{F})$-module structure. Indeed, (33) shows that if $\mathcal{G}$ is any $\mathcal{D}(\mathcal{F})$-module, then $\Gamma_{[V]}(\mathcal{G})$ is a $\mathcal{D}(\mathcal{F})$-module. We can show that the category of $\mathcal{D}(\mathcal{F})$ has enough injectives [23] so it follows, by the usual abstract yoga, that $\mathcal{H}^p_{[V]}(\mathcal{F})$ is a $\mathcal{D}(\mathcal{F})$-module. Put in such a way, this structure is obscure! But we shall give explicit representatives for sections of $\mathcal{H}^p_{[V]}(\mathcal{F})$ in a moment and then it will be quite evident.

To recover elementary states we ought to take global sections or cohomology of these local objects. Abstractly this gives

$$\mathbf{R}\Gamma_{[V]}(G/R, \cdot) = \mathbf{R}\Gamma(G/R, \mathbf{R}\underline{\Gamma}_{[V]}(\cdot)) = \varinjlim \mathbf{R}\mathsf{Hom}_{G/R}(\mathcal{O}_{G/R}/\mathcal{I}^\ell, \cdot)$$

by composition of derived functors. In other words, there is an E_2 spectral sequence

$$E_2^{p,q} = H^p(G/R, \mathcal{H}_{[V]}^q(\mathcal{F})) \implies H_{[V]}^{p+q}(G/R, \mathcal{F})$$

which allows us to compute (algebraic) relative cohomology from the cohomology of local cohomology sheaves. In the *holomorphic* category, we obtain a $\mathcal{U}(\mathbf{g})$-submodule of this relative cohomology by the natural maps

$$\varphi_i : H_{[V]}^i(G/R, \mathcal{F}_{\text{alg}}) \to H_{[V]}^i(G/R, \mathcal{F}_{\text{analytic}}).$$

Remark (11.4.5).   [$(\mathbf{g}, K)$-module structure]   When we take global cohomology and compute $H_{[V]}^{p+q}(G/R, \mathcal{F})$, using this spectral sequence, for example, we obtain modules over $\Gamma(G/R, \mathcal{D}(\mathcal{F}))$. When $\mathcal{F}$ is homogeneous, there is a natural surjection

$$\mathcal{O}_{G/R} \otimes_{\mathbf{C}} \mathcal{U}(\mathbf{g}) \to \mathcal{D}(\mathcal{F})$$

by realizing $\mathbf{g}$ as right invariant vector fields on G and sections of $\mathcal{F}$ as F-valued functions on G. So there is a mapping $\mathcal{U}(\mathbf{g}) \to \Gamma(G/R, \mathcal{D}(\mathcal{F}))$ which makes $H_{[V]}^{p+q}(G/R, \mathcal{F})$ a $\mathbf{g}$-module. This is the $\mathbf{g}$-module structure we observed in our example of elementary states above.

If, furthermore, V is a union of K orbits, then $\mathcal{I}$ is a K-homogeneous subsheaf of $\mathcal{O}_{G/R}$. It follows that $H_{[V]}^*(G/R, \mathcal{F})$ is also a K-module, compatibly with its $\mathbf{g}$-module structure. In other words, it is a Harish Chandra module.

$H_{[V]}^{p+q}(G/R, \mathcal{F})$ is related to cohomology over $Z = G/R \setminus V$ by the long exact sequence on relative cohomology:

$$\to H_V^{p-1}(G/R, \mathcal{F}) \to H^{p-1}(G/R, \mathcal{F}) \to H^{p-1}(Z, \mathcal{F}) \xrightarrow{\delta} H_V^p(G/R, \mathcal{F}) \to .$$
$$(34)$$

The groups $H^i(G/R, \mathcal{F})$ are finite dimensional, non-zero in at most one degree if $\mathcal{F}$ is irreducible and this is a sequence of $\mathcal{U}(\mathbf{g})$-modules. So we may use relative cohomology to determine $H^*(Z, \mathcal{F})$ up to a finite dimensional error (related to the need to take "reduced" cohomology at (26)). Notice that this long exact sequence and the Penrose transform (and Serre's "GAGA") show that the maps φ are injections.

We shall see that when V is smooth local cohomology occurs only in degree $p = \text{codim } V$. So then the spectral sequence becomes simply a set of equalities:

$$
\begin{aligned}
H^{i+p}_{[V]}(G/R, \mathcal{F}) &\equiv \varinjlim \text{Ext}^{i+p}_{G/R}(\mathcal{O}_{G/R}/\mathcal{I}^\ell, \mathcal{F}) \\
&= \varinjlim H^i(X, \mathcal{E}xt^p_{\mathcal{O}_{G/R}}(\mathcal{O}_{G/R}/\mathcal{I}^\ell, \mathcal{F})).
\end{aligned}
\tag{35}
$$

This is the form we shall need for computing K-types. To do that, we need to make rather more concrete our notion of local cohomology. We want explicit representatives in terms of functions with poles, etc., which look like our elementary states on which we can compute the action of K. To find these, at least when V is smooth, we need to use a *Koszul complex*.

Koszul complexes and local cohomology

Suppose that $V \overset{\iota}{\hookrightarrow} G/R$ is a smooth subvariety, codimension p, defined by the vanishing of a section s of a rank p vector bundle E. Our first task is to compute $\mathcal{E}xt^*_{\mathcal{O}_{G/R}}(\mathcal{O}_{G/R}/\mathcal{I}, \mathcal{F})$, where $\mathcal{O}_{G/R}/\mathcal{I} \cong \mathcal{O}_V$. There is a standard way of doing this, using the *Koszul complex* which resolves $\mathcal{O}_V$ as a module over $\mathcal{O}_{G/R}$ [74]:

$$
0 \to \wedge^p E^* \overset{\langle \cdot, s \rangle}{\to} \wedge^{p-1} E^* \overset{\langle \cdot, s \rangle}{\to} \cdots \to E^* \overset{\langle \cdot, s \rangle}{\to} \mathcal{O}_{G/R} \to \mathcal{O}_V \to 0.
$$

Apply $\mathcal{H}om_{\mathcal{O}_{G/R}}(\cdot, \mathcal{F})$ and take cohomology to compute $\mathcal{E}xt^*_{\mathcal{O}_{G/R}}(\mathcal{O}_V, \mathcal{F})$. Because V is smooth, this is non-zero in degree p, only:

$$
\begin{aligned}
\mathcal{E}xt^p_{\mathcal{O}_{G/R}}(\mathcal{O}_V, \mathcal{F}) &\cong \frac{\mathcal{H}om_{\mathcal{O}_{G/R}}(\wedge^p E^*, \mathcal{F})}{\text{im } \mathcal{H}om_{\mathcal{O}_{G/R}}(\wedge^{p-1} E^*, \mathcal{F}) \to \mathcal{H}om_{\mathcal{O}_{G/R}}(\wedge^p E^*, \mathcal{F})} \\
&\cong \iota^*(\det E \otimes \mathcal{F}) \\
&\cong \det N \otimes \iota^* \mathcal{F}
\end{aligned}
\tag{36}
$$

where $N \cong \iota^* E$ is the normal bundle of V in G/R.

Our second task is to compute $\mathcal{E}xt^*_{\mathcal{O}_{G/R}}(\mathcal{O}_{G/R}/\mathcal{I}^{\ell+1}, \mathcal{F})$. To do this, we employ the short exact sequences:

$$
0 \to \mathcal{I}^\ell/\mathcal{I}^{\ell+1} \to \mathcal{O}_{(\ell)} \to \mathcal{O}_{(\ell-1)} \to 0.
\tag{37}
$$

Now it is a standard fact that $\mathcal{I}^\ell/\mathcal{I}^{\ell+1} \cong \mathcal{O}_V \otimes \odot^\ell E^*$ so that

$$
\begin{aligned}
\mathcal{E}xt^p_{\mathcal{O}_{G/R}}(\mathcal{I}^\ell/\mathcal{I}^{\ell+1}, \mathcal{F}) &\cong \mathcal{E}xt^p_{\mathcal{O}_{G/R}}(\mathcal{O}_V, \odot^\ell E \otimes \mathcal{F}) \\
&\cong \iota^*(\det E \otimes \odot^\ell E \otimes \mathcal{F}) \\
&\cong \det N \otimes \odot^\ell N \otimes \iota^* \mathcal{F}
\end{aligned}
\tag{38}
$$

and vanishes in other degrees. The $\mathcal{E}xt$ long exact sequences deduced from (37) degenerate into short exact sequences:

$$0 \to \mathcal{E}xt^p_{\mathcal{O}_{G/R}}(\mathcal{O}_{(\ell-1)}, \mathcal{F}) \to \mathcal{E}xt^p_{\mathcal{O}_{G/R}}(\mathcal{O}_{(\ell)}, \mathcal{F}) \to \mathcal{E}xt^p_{\mathcal{O}_{G/R}}(\mathcal{I}^\ell/\mathcal{I}^{\ell+1}, \mathcal{F}) \to 0$$
(39)

which give a filtration of $\mathcal{H}^p_{[V]}(\mathcal{F})$ (by normal order) and show it to be zero in other degrees.

Global cohomology is obtained by taking the long exact sequence in cohomology of (39).

Example (11.4.6).   As a simple first example of the above, let $G/R = G/B = \mathbf{CP}^1$ and let $V = eB$ be a base point. Recall that elements of B are SL(2,$\mathbf{C}$) matrices of the form

$$\begin{pmatrix} * & * \\ 0 & * \end{pmatrix}$$

and let $T = K$ consist of diagonal matrices such as

$$H_\mu = \begin{pmatrix} \mu & 0 \\ 0 & 1/\mu \end{pmatrix}.$$

We calculate $H^i_{[V]}(\mathbf{CP}^1, \mathcal{O}(-n-1))$ as a T-module as follows. First, T fixes $V = eB$ and so acts on the normal bundle of V. This is the fibre of the tangent bundle Θ at eB, and so it is just the T-module

$$N = F_{\mathrm{b}}(\overset{2}{\times}).$$

(In our notation, remember, this is the T-module in which H_μ acts by multiplication by μ^{-2}.) Of course, N is one dimensional so $\det N = N$ and, therefore,

$$\mathcal{E}xt^1_{\mathcal{O}_V}(\mathcal{I}^\ell/\mathcal{I}^{\ell+1}, \mathcal{O}(-n-1)) \cong \det N \otimes \odot^\ell N \otimes \iota^* \mathcal{O}(-n-1)$$
$$\cong F_{\mathrm{b}}(\overset{2\ell-n+1}{\times})$$

as a T-module.

These $\mathcal{E}xt$ sheaves, being supported at a single point, have only sections and these are very easy to compute! Coupling the results for various ℓ together, using (39), gives

$$H^1_{[V]}(\mathbf{CP}^1, \mathcal{O}(-n-1)) \cong \varinjlim \mathrm{Ext}^1_{\mathcal{O}_V}(\mathcal{O}/\mathcal{I}^\ell, \mathcal{O}(-n-1)) \cong \bigoplus_{\ell \geq 0} F_{\mathrm{b}}(\overset{2\ell-n+1}{\times})$$
(40)

as a T-module. As a **g**-module it is the Verma module of highest weight $\overset{n-1}{\times}$ which follows easily from remark 11.1.1.

Remark (11.4.7).   We can see all of this from the "elementary state" point of view rather easily. Let $\pi_{A'}$ be homogeneous coordinates on $\mathbf{CP}^1$ and choose a basis $\{o^{A'}, \iota^{A'}\}$ of $\mathbf{C}^2$ so that eB is given by $\iota^{A'}\pi_{A'} = 0$. Realize $gl(2, \mathbf{C})$ by the following vector fields:

$$\begin{pmatrix} \dfrac{o^{A'}\partial}{\partial o_{A'}} & \dfrac{o^{A'}\partial}{\partial \iota_{A'}} \\[3mm] \dfrac{\iota^{A'}\partial}{\partial o_{A'}} & \dfrac{\iota^{A'}\partial}{\partial \iota_{A'}} \end{pmatrix}. \tag{41}$$

These correspond to the left action of $gl(2, \mathbf{C})$ on homogeneous line bundles over $\mathbf{CP}^1$; bearing in mind that this is given by

$$g \cdot f = L^*_{g^{-1}}$$

we have that the generator of $\mathbf{t}$

$$\begin{pmatrix} 1 & 0 \\ 0 & -1 \end{pmatrix} \quad \text{corresponds to} \quad -\frac{o^{A'}\partial}{\partial o_{A'}} + \frac{\iota^{A'}\partial}{\partial \iota_{A'}}.$$

The ideal $\mathcal{I}$ defining eB is generated by $\frac{(o^{A'}\pi_{A'})}{(\iota^{A'}\pi_{A'})}$ so it follows from (36), (38) that the ℓth summand in (40) is spanned by

$$\phi_\ell = \frac{(o^{A'}\pi_{A'})^{-n+\ell}}{(\iota^{A'}\pi_{A'})^{\ell+1}}.$$

Now we check the $\mathbf{t}$-weight of the summand:

$$\left(-\frac{o^{A'}\partial}{\partial o_{A'}} + \frac{\iota^{A'}\partial}{\partial \iota_{A'}} \right) \phi_\ell = (n - 1 + 2\ell)\phi_\ell.$$

This agrees with the answer above, bearing in mind the sign introduced in our notation.

We may, if we wish, switch V to be the point V' given by $o^{A'}\pi_{A'} = 0$. This is the coset

$$gB = \begin{pmatrix} 0 & 1 \\ 1 & 0 \end{pmatrix} B.$$

The above calculation can now be repeated, with the proviso that the vector fields in (41) represent elements of $gl(2, \mathbf{C})$ conjugated by g. In particular,

$$\begin{pmatrix} 1 & 0 \\ 0 & -1 \end{pmatrix} \quad \text{corresponds to} \quad \frac{o^{A'}\partial}{\partial o_{A'}} - \frac{\iota^{A'}\partial}{\partial \iota_{A'}}$$

so that

$$H^1_{[V']}(\mathbf{CP}^1, \mathcal{O}(-n-1)) \cong \bigoplus_{\ell \geq 0} F_{\mathbf{b}}\left(\begin{smallmatrix} -2\ell+n-1 \\ \times \end{smallmatrix} \right).$$

For $n \leq 0$, these are just the "K-types" of the discrete series representations $\mathcal{H}_{-n}$ encountered in section 10.1.

Example (11.4.8).   [Minkowski space]   As a second illustration, let x be the base point eP in Minkowski space and $L_x \subset Z$ the corresponding line in twistor space. We shall compute

$$H^2_{[L_x]}(\mathcal{O}(n-2)). \tag{42}$$

Actually, the simplest way of doing this whilst keeping track of **g**-module structures is to consider *all* x at once! Namely, consider the embedding

$$\times\!\!-\!\!\times\!\!-\!\!\bullet \hookrightarrow \times\!\!-\!\!\bullet\!\!-\!\!\bullet \times \bullet\!\!-\!\!\times\!\!-\!\!\bullet$$

and apply the above theory to compute local cohomology using sheaves supported on $\times\!\!-\!\!\times\!\!-\!\!\bullet$. A moment's thought shows that (42) is obtained by taking direct images to $\bullet\!\!-\!\!\times\!\!-\!\!\bullet$ and examining the stalk of the result at x.

Then

$$E = \overset{1\ \ 0\ \ 1}{\times\!\!-\!\!\times\!\!-\!\!\bullet}; \quad \det E = \overset{2\ \ 1\ \ 0}{\times\!\!-\!\!\times\!\!-\!\!\bullet}; \quad \text{and} \quad \odot^\ell E = \overset{\ell\ \ 0\ \ \ell}{\times\!\!-\!\!\times\!\!-\!\!\bullet}$$

so that with $\mathcal{F} = \mathcal{O}(n-2)$

$$\mathcal{E}xt^p_{\mathcal{O}_V}(\mathcal{I}^\ell/\mathcal{I}^{\ell+1}, \mathcal{F}) \cong \overset{n+\ell\ \ 1\ \ \ell}{\times\!\!-\!\!\times\!\!-\!\!\bullet}.$$

To use (35) take zeroth direct image τ_* of this, obtaining

$$\overset{n+\ell\ \ 1\ \ \ell}{\bullet\!\!-\!\!\times\!\!-\!\!\bullet} \qquad \text{provided } n+\ell, \ell \geq 0.$$

Restriction to x now amounts to viewing these as finite dimensional K-modules. Then $H^2_{[L_x]}(\mathbf{CP}^3, \mathcal{O}(n-2))$ is their direct sum, as a K-module.

Explicit hyperfunction representatives

We want now to show directly that the abstract calculations given above have a simple concrete realization. In this we can compute explicit local representatives of $\mathcal{H}^p_{[V]}(\mathcal{F})$ in terms of *hyperfunctions* or meromorphic sections of $\mathcal{F}$. We will suppose that $\mathcal{F}$ is a homogeneous sheaf but the construction is valid more generally. It depends on the following two lemmata (the first of which is a variant of Kashiwara's theorem).

Lemma (11.4.9).   If $\iota: V \to G/R$ is a smooth embedding of codimension p, then $\mathcal{H}^p_{[V]}(\mathcal{F})$ is generated over the sheaf $\mathcal{D}(\mathcal{F})$ of differential operators on $\mathcal{F}$ by $\mathcal{E}xt^p_{\mathcal{O}_{G/R}}(\mathcal{O}_V, \mathcal{F})$.

Proof   [23, p. 261] The category of $\mathcal{D}(\mathcal{F})$-modules has enough injectives [23]. So, assuming $\mathcal{F}$ injective, it therefore suffices to show that $_{\mathcal{I}^\ell}\mathcal{F}$ is generated over $\mathcal{D}(\mathcal{F})$ by $_\mathcal{I}\mathcal{F}$. Proceed inductively, supposing the result true for ℓ and letting $v \in {}_{\mathcal{I}^\ell}\mathcal{F}$. Let $f = (f_1, \ldots, f_p)$ define V locally, so that the monomials f^ν (ν a multi-index with $|\nu| = \ell$) generate $\mathcal{I}^\ell$ over $\mathcal{O}_{G/R}$. Then

$$f^\nu \left\{ \ell v + \sum \frac{\partial}{\partial f_i} f_i v \right\} = \ell f^\nu v + \sum \left\{ \frac{\partial}{\partial f_i} f^\nu f_i - \nu_i f^\nu \right\} v$$
$$= 0.$$

So, by induction,

$$\ell v + \sum \frac{\partial}{\partial f_i} f_i v \in \mathcal{D}(\mathcal{F})_\mathcal{I}\mathcal{F}.$$

But then $f_i v \in {}_{\mathcal{I}^\ell}\mathcal{F}$, so $f_i v$ and hence $\frac{\partial}{\partial f_i} f_i v \in \mathcal{D}(\mathcal{F})_\mathcal{I}\mathcal{F}$.   □

Now suppose that $\mathcal{F}$ is homogeneous. Consider the subsheaf of $\mathcal{D}(\mathcal{F})$ consisting of functions taking values in $\mathcal{U}(1_x)$ over $\mathbf{r}_x \in G/R$ with reductive Levi factor 1_x) and take its image in $\mathcal{D}(\mathcal{F})$. Then it is clear that any non-zero section v of $\mathcal{F}$ generates $\mathcal{F}$ locally under the action of $\mathcal{D}(\mathcal{F})$. Similarly, $v_{|V}$ generates $\iota^*\mathcal{F}$ locally over $\mathcal{D}(\iota^*\mathcal{F})$. So, by the theorem,

$$v \frac{\partial}{\partial f_1} \wedge \ldots \wedge \frac{\partial}{\partial f_p}$$

generates $\mathcal{E}xt^p_{\mathcal{O}_{G/R}}(\mathcal{O}_V, \mathcal{F})$ over $\mathcal{D}(\iota^*\mathcal{F})$ and we have proved:

Lemma (11.4.10).   *Let $\iota : V \to G/R$ be a smooth embedding, locally defined by the vanishing of co-ordinates $f_1 \ldots f_p$, and $\mathcal{F}$ an irreducible homogeneous sheaf on G/R. If v is any local section of $\mathcal{F}$, then $v\frac{\partial}{\partial f_1} \wedge \ldots \wedge \frac{\partial}{\partial f_p}$ generates $\mathcal{H}^p_{[V]}(\mathcal{F})$ locally, over $\mathcal{D}(\mathcal{F})$.*

(For a fuller understanding of the $\mathcal{D}(\mathcal{F})$-module theory underlying this result see [23, VI, 7.9 & 7.10], where $\underline{\Gamma}_{[V]}$ is interpreted in the category of $\mathcal{D}(\mathcal{F})$ modules.)

Now, following Schapira [132], we may turn the wheel of abstraction full circle to regain contact with elementary states. Define, in the local situation of the lemma,

$$\mathcal{K}(\mathcal{F}) = \left\{ \sum_{0 < |\nu| \le m} a_\nu f^{-\nu} \mid m \in \mathbf{N}, a_\nu \text{ is a local section of } \iota^*\mathcal{F}, \nu \in \mathbf{N}^p \right\}$$

and

$$\mathcal{L}(\mathcal{F}) = \left\{ \sum_{0 < |\nu| \leq m \text{ and } \nu_i = 0 \text{ at least once}} a_\nu f^{-\nu} \right\}.$$

Then

$$\delta_f(v) = \frac{v}{f_1 \cdots f_p}$$

generates $\mathcal{K}(\mathcal{F})/\mathcal{L}(\mathcal{F})$ under the obvious action of $\mathcal{D}(\mathcal{F})$. Its annihilator is the left ideal generated by $\mathcal{I}$ and the annihilator of v in $\mathcal{D}(\iota^*\mathcal{F})$ (identified locally with the commutator of multiplication by $\mathcal{I}$). But this is precisely the annihilator of $v\dfrac{\partial}{\partial f_1} \wedge \ldots \wedge \dfrac{\partial}{\partial f_p}$. In other words, the correspondence

$$\frac{v}{f_1 \cdots f_p} \longleftrightarrow v\frac{\partial}{\partial f_1} \wedge \ldots \wedge \frac{\partial}{\partial f_p}$$

establishes a local isomorphism of $\mathcal{D}(\mathcal{F})$-modules

$$\mathcal{K}(\mathcal{F})/\mathcal{L}(\mathcal{F}) \cong \mathcal{H}^p_{[V]}(\mathcal{F})$$

which is easily checked to be invariant under a change of defining coordinates for V. In particular the left-hand side makes global sense and defines a sheaf which is often called the sheaf of *hyperfunctions* on V with values in $\mathcal{F}$.

Example (11.4.11).   Let us illustrate the previous remark in the context of the twistorial elementary states given at the start of this section. The point given by

$$A_\alpha Z^\alpha = B_\alpha Z^\alpha = C_\alpha Z^\alpha = 0 \text{ and } D_\alpha Z^\alpha \neq 0$$

lies on **L** and

$$s = \frac{A_\alpha Z^\alpha}{D_\alpha Z^\alpha} \qquad t = \frac{B_\alpha Z^\alpha}{D_\alpha Z^\alpha}$$

are locally good normal co-ordinates to **L**. Elementary states are finite linear combinations of terms such as

$$\frac{a}{s^i t^j} \quad \text{for } a \text{ a local section of } \iota^* \overset{-n-2 \quad 0 \quad 0}{\times\!\!-\!\!\bullet\!\!-\!\!\bullet}$$

hence sections of $\mathcal{H}^p_{[\mathbf{L}]}(\overset{-n-2 \quad 0 \quad 0}{\times\!\!-\!\!\bullet\!\!-\!\!\bullet})$. Indeed, they are precisely the *global* sections of this sheaf. The local $\mathcal{D}(\overset{-n-2 \quad 0 \quad 0}{\times\!\!-\!\!\bullet\!\!-\!\!\bullet})$-module structure induces the global action of $\mathbf{sl}(4, \mathbf{C})$.

K-types and Grothendieck duality

Suppose, as before, that P has a reductive Levi factor K; this may be the complexification of a maximal compact subgroup of a real form G_u of G. We shall suppose that an opposite parabolic P^t has been chosen so that $P \cap P^t = K$. We now want to use the machinery just developed to evaluate K-types in the cohomology groups $H^*(Z, \mathcal{F})$ where $\mathcal{F}$ is an irreducible homogeneous sheaf on G/R and $Z \subset G/R$ is either $P^t.eR$ or $G_u.eR$ or its closure.

The easiest situation is when $V = G/R \setminus Z$ is a smooth closed sub-variety of G/R and, therefore, a union of K-orbits. For then we can use the Koszul complex of V as above to calculate the relative cohomology $H^{*+1}_{[V]}(G/R, \mathcal{F})$ as a K-module and substitute this in the long exact sequence on relative cohomology (34).

Example (11.4.12).   [Discrete series for SU(1, 1)]   From example 11.4.6, we have that

$$H^1_{[V']}(\mathbf{CP}^1, \mathcal{O}(-n-1)) \cong \bigoplus_{\ell \geq 0} F_{\mathbf{b}}(\begin{smallmatrix} -2\ell+n-1 \\ \times \end{smallmatrix}).$$

This and the long exact sequence on relative cohomology (in the *algebraic category*):

$$0 \to \Gamma(\mathbf{CP}^1, \mathcal{O}(-n-1)) \to \Gamma(Z, \mathcal{O}(-n-1)) \to H^1_{[V']}(\mathbf{CP}^1, \mathcal{O}(-n-1)) \to 0$$

computes the K-types of H_{-n}. It is easy to see that a holomorphic section of $\mathcal{O}(-n-1)$ over Z is K-finite only if it is algebraic, i.e., only if it is a linear combination of the ϕ_ℓ given in 11.4.6.

Example (11.4.13).   [Ladder series for SU(2, 2)] Consider example 11.4.8. The long exact sequence in relative cohomology gives, with coefficients in $\mathcal{O}(-n-2)$ (in the algebraic category, again)

$$\begin{array}{ccccccc} H^1(\mathbf{CP}^3) & \to & H^1(Z) & \stackrel{\sim}{\to} & H^2_{[\mathbf{L}]}(\mathbf{CP}^3) & \to & H^2(\mathbf{CP}^3). \\ \| & & & & & & \| \\ 0 & & & & & & 0 \end{array}$$

To compute cohomology relative to $\mathbf{L}$ from cohomology relative to L_x (given in 11.4.8), we should conjugate by the following element of SL(4,$\mathbf{C}$) which interchanges L_x and $\mathbf{L}$:

$$\begin{pmatrix} 0 & 0 & 1 & 0 \\ 0 & 0 & 0 & 1 \\ 1 & 0 & 0 & 0 \\ 0 & 1 & 0 & 0 \end{pmatrix}.$$

This sends the weight $\overset{p\ \ q\ \ r}{\bullet\!-\!\bullet\!-\!\bullet}$ to $\overset{r\ p+q+r\ p}{\bullet\!-\!\bullet\!-\!\bullet}$ and corresponds to the Weyl group element $\sigma_2\sigma_1\sigma_3\sigma_2$ which is the longest element of $W^{\mathbf{P}}$ where $\mathbf{p} = \bullet\!\!-\!\!\!\times\!\!\!-\!\!\bullet$. So as a K-module

$$H^1(Z, \mathcal{O}(-n-2)_{\mathrm{alg}}) \cong \bigoplus_{\ell \geq \max\{0,n\}} F_{\mathbf{p}}(\ \overset{\ell\ \ \ n\text{-}2\ell\text{-}1\ \ \ell\text{-}n}{\bullet\!\!-\!\!\!\times\!\!\!-\!\!\bullet}\ )$$

(see (24)). It remains to see that these are *all* the K-types in the holomorphic cohomology on Z, even when Z is taken to be the closure of an orbit of $\mathrm{SU}(2,2)$ like $\overline{\mathbf{PT}^+}$. That will be seen from the arguments to follow.

This is all very well, provided V is a smooth closed subvariety of G/R. It almost never is. Usually it is either a singular Schubert subvariety of G/R (when $Z = P^t.eR$) or a holomorphic submanifold, with a possibly singular boundary (when Z is (the closure of) a union of G_u orbits).

For example, in the ambitwistor double fibration,

$$Z = P^t.eR \subset \overset{\times\!-\!\bullet\!-\!\times}{} \qquad \overset{\times\!-\!\times\!-\!\times}{\underset{\eta\ \nearrow \qquad\qquad \searrow\ \tau}{}} \qquad \overset{\bullet\!-\!\times\!-\!\bullet}{} \supset X = P^t.eP$$

the complement V of Z is a pair of $\mathbf{CP}^3$'s each blown up along a common quadric. (Physically, this corresponds to the light rays which lie in the light cone of the point "at infinity" for X. These are intersections of α- or β-planes with the light cone (giving the blown-up $\mathbf{CP}^3$'s); the generators of the cone lie in both families and form the common quadric.)

Actually, there is still a way to use Koszul complexes, etc., to compute K-types (for good cases) provided we shift our point of view slightly.

Let $\eta(\tau^{-1}x) = L_x \subset Z$ for $x = eP \in G/P$. Then $L_x = K.eR \cong K/K \cap R$ so it certainly is a smooth (homogeneous) subvariety of Z. We can use the methods of the previous section to compute, say,

$$H^d_{[L_x]}(G/R, \mathcal{F}^* \otimes \Omega^n) \cong \varinjlim \mathrm{Ext}^d_{\mathcal{O}_{G/R}}(\mathcal{F} \otimes_{\mathcal{O}} \mathcal{O}/\mathcal{I}^\ell, \Omega^n).$$

Here $n = \dim G/R$ and $\mathcal{I}$ defines L_x. The point is that each term in the limit on the right of this equation has a natural dual, which is easy to construct geometrically:

Theorem (11.4.14). [Grothendieck duality] *There is a perfect pairing of K-modules*

$$\mathsf{Ext}^d(\mathcal{F}\otimes\mathcal{O}/\mathcal{I}^\ell,\Omega^n)\times H^{n-d}(G/R,\mathcal{F}\otimes\mathcal{O}/\mathcal{I}^\ell)\to H^n(G/P,\Omega^n)\cong\mathbf{C}.$$

(For a detailed explanation of the duality theorem of Grothendieck on which this is based see [74]. We can use this to give an abstract account of Penrose's contour integral formulae which amounts to evaluating residues along L_x as x varies through X.)

Of course, there is a natural family of maps

$$\rho_\ell : H^{n-d}(Z,\mathcal{F})\to H^{n-d}(G/R,\mathcal{F}\otimes\mathcal{O}/\mathcal{I}^{\ell+1})$$

consistent under further restriction. So deduce a map

$$\rho : H^{n-d}(Z,\mathcal{F})\to\varprojlim H^{n-d}(G/R,\mathcal{F}\otimes\mathcal{O}/\mathcal{I}^{\ell+1}).$$

Under the Penrose transform ρ_ℓ becomes restriction to the ℓth formal neighbourhood of x in G/P; so by elementary complex analysis we have:

Lemma (11.4.15). ρ *is injective.*

Putting this together with theorem 11.4.14 yields

Theorem (11.4.16). *The K-types of $H^{n-d}(Z,\mathcal{F})$ occur amongst the contragredients of the K-invariant direct summands of*

$$H^d_{[Z_x]}(G/R,\mathcal{F}^*\otimes\Omega^n).$$

Actually, as the reader will readily check, it is often the case that the K-types are exactly given by $H^{d-s}_{[Z_x]}(G/R,\mathcal{F}^*\otimes\Omega^d)$. Consider, for example, the ladder series for $SU(2,2)$ and the corresponding series for Hermitian symmetric spaces. What is required is that the maps ρ_ℓ should be surjective. We shall not study general conditions for this here but merely remark on two cases in which they are.

The first is when the Penrose transform of $H^s(Z,\mathcal{F})$ is the space of solutions of a system of equations and not the subquotient of a second differential operator. For then it may be possible to extend each finite order solution of the equations to a solution on X, in which case ρ_ℓ is surjective.

The second possibility is that there may exist a second K-orbit $L_x^\perp$ with

$$Z\subset G/R\backslash L_x^\perp$$

so that we have maps

$$H^*(G/R\backslash L_x^\perp,\mathcal{F})\to H^*(Z,\mathcal{F})\to\varprojlim H^*(G/R,\mathcal{F}\otimes\mathcal{O}/\mathcal{I}^{\ell+1}).$$

We may then compute K-types in the first of these spaces, using the methods outlined below and compare them with those in the last, just computed using Grothendieck duality. If they agree, then they agree with the K-types in $H^*(Z, \mathcal{F})$.

11.5   Homomorphisms of Verma modules

In theorem 11.2.1, homomorphisms of (generalized) Verma modules and invariant linear differential operators between homogeneous sheaves were identified. As observed there, any differential operator occurring in the Penrose transform is evidently invariant and so the Penrose transform gives a method of constructing homomorphisms of Verma modules.

As remarked in section 11.1, homomorphisms of generalized Verma modules (for some parabolic $\mathbf{p}$) are of two kinds. The simplest are the *standard* homomorphisms, which correspond to invariant differential operators on G/P which are direct images of operators on G/B for B, a Borel subgroup. When such a direct image is zero, there may nonetheless exist a non-zero invariant differential operator; the corresponding homomorphism of generalized Verma modules is not covered by a homomorphism of Verma modules and is called *non-standard* [110]. These non-standard operators or homomorphisms are obtained in the Penrose transform whenever the hypercohomology spectral sequence takes more than one step to converge. Put another way, the standard operators are obtained if the (derived) direct image functor is applied to a complex of length two (whose differential is the inducing operator on G/B); applying it to longer complexes will produce non-standard operators.

The construction is best illustrated by means of an example. Let

$$\mathbf{p} = \; \times\!\!-\!\!\bullet\!\!-\!\!\bullet \; \cdots \; \bullet\!\!-\!\!\bullet\!\!\!<^{\textstyle\bullet}_{\textstyle\bullet}$$

$$\mathbf{r} = \; \bullet\!\!-\!\!\bullet\!\!-\!\!\bullet \; \cdots \; \bullet\!\!-\!\!\bullet\!\!\!<^{\textstyle\times}_{\textstyle\bullet}$$

and consider the Penrose transform based on the double fibration

$$G/R \xleftarrow{\eta} G/Q \xrightarrow{\tau} G/P.$$

We shall construct non-standard homomorphisms of generalized Verma modules for $\mathbf{p}$ which are of non-singular infinitesimal character.

In order to do this, we shall vary the subject of the transform through all homogeneous sheaves $\mathcal{O}_{\mathbf{r}}(w.\lambda)$ where $w \in W^{\mathbf{r}}$ and λ is dominant for $\mathbf{g}$.

Indeed, by the translation principle, we may take $\lambda = 0$. The first task is therefore to compute $W^{\mathbf{r}}$ using the method of section 4.4. Denote by σ_i the simple reflection corresponding to the ith node in the Dynkin diagram for **g**, with simple roots labelled from left to right (and top to bottom). Observe that

$$\rho^{\mathbf{r}} = \overset{0}{\underset{\times}{}} \; \overset{0}{\bullet} \; \overset{0}{\bullet} \; \cdots \; \overset{0}{\bullet} \; \overset{0}{\bullet} \overset{\displaystyle 1}{\diagup} \; .$$

Then the only $w \in W^{\mathbf{r}}$ of length one is σ_n and of length two is $\sigma_n \sigma_{n-1}$. Thereafter, there is much choice. The most basic path in $W^{\mathbf{r}}$ consists of elements of the form

$$t_k = \sigma_n \sigma_{n-1} \sigma_{n-2} \cdots \sigma_{n-k+1} \quad \text{for} \quad k = 0, 1, \ldots, n-1.$$

From this path, several shorter parallel paths are derived whose elements contain σ_{n+1}; for example, the next path is given by

$$u_k = t_k \sigma_{n+1} \quad \text{for} \quad k = 2, 3, \ldots, n-1.$$

These paths yield an initial block in $W^{\mathbf{r}}$ as shown in figure 11.1.

   G/R is, recall, a space of (reduced) projective pure spinors for $\mathrm{SO}(n+2,\mathbf{C})$. Choose an embedding $\mathbf{so}(n,\mathbf{C}) \subset \mathbf{so}(n+2,\mathbf{C})$ by letting the simple roots for $\mathbf{so}(n,\mathbf{C})$ correspond to $\alpha_2, \alpha_3, \ldots, \alpha_{n+1}$; this realizes the space of projective pure spinors for $\mathbf{so}(n,\mathbf{C})$ as a smooth subvariety of G/R of codimension n. (It is the subvariety corresponding under the double fibration to an origin stabilized by $\mathrm{SO}(n,\mathbf{C})$ in G/P.) G/R may be built from this variety by attaching various Bruhat cells which are parametrized by the elements of the initial block of $W^{\mathbf{r}}$ just constructed. The remaining elements of $W^{\mathbf{r}}$ correspond to Bruhat cells building the subvariety. This is, in turn, a variety of projective pure spinors, and the construction of $W^{\mathbf{r}}$ may now proceed inductively; the Hasse diagram for case $\mathbf{so}(n+2,\mathbf{C})$ is obtained from that in case $\mathbf{so}(n,\mathbf{C})$ by attaching an initial block. The result is given in figure 11.2.

Remark (11.5.1).   This inductive construction of pure spinors lifts to an inductive construction of spinors which is beautifully described in the appendix of [128].

   We shall be most interested in those elements of $W^{\mathbf{r}}$ which begin each successive block. The first, of course, is $w_0 = \mathrm{id}$. The second lies at the end of the basic path of t_i's and is therefore

$$w_1 = \sigma_n \sigma_{n-1} \sigma_{n-2} \cdots \sigma_1.$$

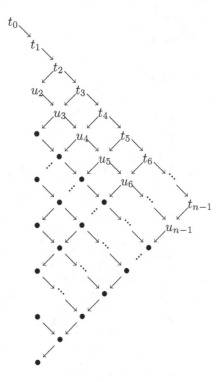

Figure 11.1. Initial block of Hasse diagram for **r**

(Observe that $\ell(w_1) = n$, in accord with the codimension of the first subvariety.) Inductively,

$$w_k = \begin{cases} w_{k-1}\sigma_{n+1}\sigma_{n-1}\sigma_{n-2}\cdots\sigma_k & k \text{ even} \\ w_{k-1}\sigma_n\sigma_{n-1}\sigma_{n-2}\cdots\sigma_k & k \text{ odd.} \end{cases}$$

These are reduced expressions, so

$$\ell(w_k) = \ell(w_{k-1}) + n - k + 1 = k(2n - k + 1)/2$$

(and, again, the relative codimensions are correct).

The next step in the transform is to compute the relative Hasse diagram $W_{\mathbf{r}}^{\mathrm{P}}$ so as to find the Bernstein–Gelfand–Gelfand resolutions along η. This is easily checked to be

$$W_{\mathbf{r}}^{\mathrm{P}} = \{\mathrm{id}, \sigma_1, \sigma_1\sigma_2, \ldots, \sigma_1\cdots\sigma_{n-1}, \sigma_1\cdots\sigma_{n-1}\sigma_{n+1}\}.$$

Denote the element of length j in this by v_j. Then the Bernstein–Gelfand–Gelfand resolution of $\eta^{-1}\mathcal{O}_{\mathbf{r}}(w_k.\lambda)$ has k^{th} term $\mathcal{O}_{\mathbf{q}}(v_j w_k.\lambda)$.

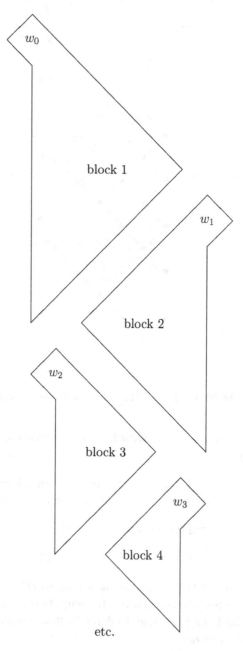

Figure 11.2.   Hasse diagram of **r**

Finally, $R^\ell \tau_* \mathcal{O}_q(v_j w_k.\lambda)$ is non-zero only when $\ell = \ell(w)$ where $v_j w_k = ww'$ with $w \in W_p^r$ (the Hasse diagram for the fibration τ) and $w' \in W^P$. The direct image is, in that case, just $\mathcal{O}_p(w'.\lambda)$. Observe that W_p^r is easily obtained from the calculation of W^r given above, since the fibre of τ is still a projective pure spinor variety. W^P is given in example 4.3.7. Carrying this out we obtain:

Lemma (11.5.2).   *Let $1 \le k \le n-1$. Then the $E_1^{p,q}$ term in the spectral sequence computing the Penrose transform of $\mathcal{O}_r(w_k.0)$ is*

$$
E_1^{p,q} \left|
\begin{array}{cccccccc}
0 & 0 & \cdots & 0 & 0 & 0 & \cdots & 0 \\
\Omega^k & \Omega^{k+1} & \cdots & \Omega^n_\pm & 0 & 0 & \cdots & 0 \\
0 & 0 & \cdots & 0 & 0 & 0 & \cdots & 0 \\
\vdots & \vdots & \cdots & \vdots & \vdots & \vdots & \cdots & \vdots \\
0 & 0 & \cdots & 0 & \Omega^{2n-k+1} & \Omega^{2n-k+2} & \cdots & \Omega^{2n} \\
0 & 0 & \cdots & 0 & 0 & 0 & \cdots & 0
\end{array}
\right.
$$

where the non-trivial rows are at $q = nk - k(k+1)/2, n(k-1) - (k-1)k/2$, and $\Omega^{2n-(k-1)}$ lies in column $n - k + 1$. The rows form part of the holomorphic de Rham sequence on G/P (and the occurrence of Ω^n_+ or Ω^n_- depends on the parity of k).

Deriving once yields

$$
E_2^{p,q} \left|
\begin{array}{cccccc}
0 & 0 & \cdots & 0 & 0 & 0 \\
\ker \Omega^k \to \Omega^{k+1} & 0 & \cdots & \mathrm{coker}\, \Omega^{n-1} \to \Omega^n_\pm & 0 & 0 \\
0 & 0 & \cdots & 0 & 0 & 0 \\
\vdots & \vdots & \cdots & \vdots & \vdots & \vdots \\
0 & 0 & \cdots & 0 & \ker \Omega^{2n-k+1} \to \Omega^{2n-k+2} & 0 \\
0 & 0 & \cdots & 0 & 0 & 0
\end{array}
\right.
$$

since the holomorphic de Rham resolution is exact when evaluated over affine Minkowski space in G/P. The following terms in the spectral sequence are stable until we reach $E_{n-k+1}^{p,q}$ when we obtain a differential:

$$d_{n-k+1} : \ker \{\Omega^k \to \Omega^{k+1}\} \longrightarrow \ker \{\Omega^{2n-k+1} \to \Omega^{2n-k+2}\}.$$

Using the exactness of the de Rham complex again, this may be written as

$$d_{n-k+1} : \mathrm{im}\, \{\Omega^{k-1} \to \Omega^k\} \longrightarrow \ker \{\Omega^{2n-k+1} \to \Omega^{2n-k+2}\}$$

which, when composed with the exterior differential $d : \Omega^{k-1} \to \Omega^k$ and writing $k - 1$ for j, gives an invariant differential operator (i.e., a non-standard homomorphism):

$$D : \Omega^j \to \Omega^{2n-j} \quad \text{for } 0 \le j \le n - 2.$$

These are *all* the non-standard homomorphisms in this case [19].

CONCLUSIONS AND OUTLOOK

To conclude, we should like to mention briefly some further directions for research, especially in representation theory. There are two major topics to consider, namely homomorphisms of Verma modules and unitary representations.

We have seen in chapter 10 how the Penrose transform, in the guise of the twistor transform, leads to the construction of the ladder series of representations for $SU(p,q)$; it is clear that similar techniques will generate ladder representations for Hermitian symmetric pairs [9]. These all occur as the holomorphic cohomology of line bundles; we should like to understand those representations which arise from applying the twistor transform to vector bundles. Not all of these will be unitary (i.e., the twistor transform will not produce a suitable isomorphism) but we expect that those which are should lie in the continuation of the holomorphic discrete series for a Hermitian symmetric pair. On the other hand, we anticipate that there is a variant of the twistor transform, defined on the full flag variety G/B, which will allow the construction of the full discrete series, for appropriate G. Preliminary calculations in [41] show that representations with the correct K-types can be constructed; by a theorem of Schmid, this means that these representations are in the discrete series. These calculations are, however, at an infinitesimal level, and are not yet refined enough to indicate how the K-types fit together in the representation. When this is understood, a self contained cohomological construction of the discrete series will have been obtained which avoids all the analytical difficulties associated to L^2-cohomology calculations. The relationship between the geometry of the flag variety and the discrete series will be manifest.

In the final section, we indicated how the Penrose transform generates (non)-standard homomorphisms of Verma modules. It is an intriguing question as to why this should be so and what class of homomorphisms can be generated. We hope to establish the relationship between this construction and the $\mathcal{D}$-module constructions of Beilinson and Bernstein.

In addition to the two main topics, there are several others that merit attention. The methods of this book extend to a setting in which G/P is replaced by a curved manifold. This has structure group K where K is a reductive Levi factor of P and (G, K) is a Hermitian symmetric pair—accordingly, these manifolds have been called "almost Hermitian symmetric" [6,10]. Using the Penrose transform, it is possible to study the deformations of such structures. We have also mentioned the possibility of a symplectic Penrose transform which may be of interest in studying symplectic manifolds with a symplectic action of a semisimple complex Lie group G; very little work has been done on that so far.

From the physicist's point of view, we feel that representation theory is more than merely a useful computational tool in twistor theory. As our understanding of the subject has developed, we have been struck by the scale to which twistor-theoretic constructions can be given representation-theoretic interpretations. We believe that this is not merely accidental, for our understanding of physics should ultimately depend on natural symmetry principles (or even combinatorics) in a geometric setting. As a specific example, we now know that zero rest mass free fields correspond to certain irreducible Verma modules. We should expect that understanding how to decompose tensor products of such modules would lead to a theory of *interacting* fields. The projections onto components of such a tensor product would be twistorial vertex operators.

REFERENCES

[1] A. Andreotti and H. Grauert. Théorèmes de finitude pour des espaces complexes. *Bull. Soc. Math. Fr.* **90** (1962), 193–259.

[2] M. F. Atiyah and F. Hirzebruch. Vector bundles and homogeneous spaces. In: Proc. Symp. Pure Math. III (Differential Geometry) (A.M.S. 1961) pp. 7–38.

[3] M. F. Atiyah, N. J. Hitchin, and I. M. Singer. Self-duality in four-dimensional Riemannian geometry. *Proc. R. Soc. London* **A362** (1978), 425–461.

[4] M. F. Atiyah and W. Schmid. A geometric construction of the discrete series for semisimple Lie groups. *Inv. Math.* **42** (1977), 1–62.

[5] T. N. Bailey, L. Ehrenpreis, and R. O. Wells, Jr. Weak solutions of the massless field equations. *Proc. R. Soc. London* **A384** (1982), 403–425.

[6] T. N. Bailey and M. G. Eastwood. Complex paraconformal manifolds, their differential geometry and twistor theory. Preprint 1988.

[7] T. N. Bailey. Twistors and fields with sources on worldlines. *Proc. R. Soc. London* **A397** (1985), 143–155.

[8] R. J. Baston. *The Algebraic Construction of Invariant Differential Operators.* D.Phil. thesis. Oxford University 1985.

[9] R. J. Baston. A cohomological construction of ladder representations for Hermitian symmetric pairs. Preprint 1989.

[10] R. J. Baston. Almost Hermitian symmetric manifolds I—local twistor theory. Preprint 1989.

[11] R. J. Baston. Almost Hermitian symmetric spaces II—the Penrose transform. Preprint 1989.

[12] R. J. Baston and L. J. Mason. Conformal gravity, the Einstein equations, and spaces of complex null geodesics. *Classical Quantum Gravity* **4** (1987), 815–826.

[13] H. Bateman. The solution of partial differential equations by means of definite integrals. *Proc. Lond. Math. Soc.* **1** (1904), 451–458.

[14] H. Bateman. The transformation of the electrodynamical equations. *Proc. L.M.S.* **8** (1910), 223–264.

[15] A. Beilinson and J. N. Bernstein. Localisation de **g**-modules. *C. R. Acad. Sci.* Paris **292** (1981), 15–18.

[16] H. J. Bernstein and A. V. Phillips. Fiber bundles and quantum theory. *Sci. Am.* **245** (1981), 94–109.

[17] I. N. Bernstein, I. M. Gelfand, and S. I. Gelfand. Differential operators on the base affine space and a study of **g**-modules. In: *Lie Groups and their Representations* (ed. I. M. Gelfand. Adam Hilger 1975) pp. 21–64.

[18] I. N. Bernstein, I. M. Gelfand, and S. I. Gelfand. Schubert cells and the cohomology of the spaces G/P. In: *Representation Theory* (L.M.S. Lecture Notes vol. 69. Cambridge University Press 1982) pp. 115–140.

[19] B. D. Boe and D. H. Collingwood. A comparison theory for the structure of induced representations *I*. *J. Alg.* **94** (1985), 511–545.

[20] B. D. Boe and D. H. Collingwood. A comparison theory for the structure of induced representations *II*. *Math. Z.* **190** (1985), 1–11.

[21] W. M. Boothby. Homogeneous complex contact manifolds. In: Proc. Symp. Pure Math. III (Differential Geometry) (A.M.S. 1961) pp. 144–154.

[22] A. Borel and F. Hirzebruch. Characteristic classes and homogeneous spaces I, II, III. *Am. J. Math.* **80** (1958), 458–538, **81** (1959), 315–382, **82** (1960), 491–504.

[23] A. Borel, et al.. *Algebraic D-Modules. Perspect. in Math.* vol. 2. Academic Press 1987.

[24] R. Bott. Homogeneous vector bundles. *Ann. Math.* **60** (1957), 203–248.

[25] R. Bott and L. W. Tu. *Differential Forms in Algebraic Topology.* Grad. Texts in Math. vol. 82. Springer 1982.

[26] R. L. Bryant. Lie groups and twistor theory. *Duke Math. J.* **52** (1985), 223–261.

[27] R. L. Bryant. Private communication.

[28] J. L. Brylinski and M. Kashiwara. Kahzdan–Lusztig conjecture and holonomic systems. *Inv. Math.* **64** (1981), 387–410.

[29] N. P. Buchdahl. *Applications of Several Complex Variables in Twistor Theory*. M. Sc. thesis. Oxford University 1980.

[30] N. P. Buchdahl. A generalized de Rham sequence. *Twistor Newsletter* **10** (1980), 11–13.

[31] N. P. Buchdahl. On the relative de Rham sequence. *Proc. A.M.S.* **87** (1983), 363–366.

[32] N. P. Buchdahl. Analysis on analytic spaces and non-self-dual Yang-Mills fields. *Trans. A.M.S.* **288** (1985), 431–469.

[33] D. Burns. Some background and examples in deformation theory. In: *Complex Manifold Techniques in Theoretical Physics* (eds. D. E. Lerner and P. D. Sommers. Research Notes in Math. vol. 32. Pitman 1979) pp. 135–153.

[34] H. Cartan and J.-P. Serre. Un théorème de finitude concernant les variétés analytiques compactes. *C. R. Acad. Sci. Paris* **237** (1953), 128–130.

[35] D. H. Collingwood. *Representations of rank one Lie groups*. Research Notes in Math. vol. 137. Pitman 1985.

[36] E. Cunningham. The principal of relativity in electrodynamics and an extension thereof. *Proc. L.M.S.* **8** (1910), 77–98.

[37] M. Demazure. A very simple proof of Bott's theorem. *Inv. Math.* **33** (1976), 271–272.

[38] P. A. M. Dirac. The quantum theory of the electron I, II. *Proc. R. Soc. London* **A117** (1928), 610–624, **A118** (1928), 351–361.

[39] J. Dixmier. *Enveloping Algebras*. North-Holland 1977.

[40] E. G. Dunne. Hyperfunctions in representation theory and mathematical physics. In: *Integral Geometry* (Proceedings of a Summer Research Conference, August 1984; eds. R.L. Bryant, V. Guillemin, S. Helgason and R.O. Wells jr. : *Contemp. Math.* **63**. A.M.S. 1987).

[41] E. G. Dunne and M. G. Eastwood. Twistor transform for discrete series. Oxford Preprint 1989.

[42] M. G. Eastwood. Some cohomological arguments applicable to twistor theory. In: *Advances in Twistor Theory* (eds L. P. Hughston and R. S. Ward. Research Notes in Math. vol. 37. Pitman 1979) pp. 72–82.

[43] M. G. Eastwood and L. P. Hughston. Massless fields based on a line. In: *Advances in Twistor Theory* (eds L. P. Hughston and R. S. Ward. Research Notes in Math. vol. 37. Pitman 1979) pp. 101–109.

[44] M. G. Eastwood, R. Penrose, and R. O. Wells, Jr.. Cohomology and massless fields. *Commun. Math. Phys.* **78** (1981), 305–351.

[45] M. G. Eastwood and M. L. Ginsberg. Duality in twistor theory. *Duke Math. J.* **48** (1981), 177–196.

[46] M. G. Eastwood and K. P. Tod. Edth—a differential operator on the sphere. *Math. Proc. Camb. Philos. Soc.* **92** (1982), 317–330.

[47] M. G. Eastwood. The generalized twistor transform and unitary representations of $SU(p,q)$. Preprint.

[48] M. G. Eastwood. On Michael Murray's twistor correspondence. *Twistor Newsl.* **19** (1985), 24.

[49] M. G. Eastwood. Complexification, twistor theory, and harmonic maps from a Riemann surface. *Bull. A.M.S.* **11** (1984), 317–328.

[50] M. G. Eastwood. The generalized Penrose-Ward transform. *Math. Proc. Camb. Philos. Soc.* **97** (1985), 165–187.

[51] M. G. Eastwood. A duality for homogeneous bundles on twistor space. *J.L.M.S.* **31** (1985), 349–356.

[52] M. G. Eastwood. Supersymmetry, twistors and the Yang-Mills equations. *Trans. A.M.S.* **301** (1987), 615–635.

[53] M. G. Eastwood and J. W. Rice. Conformally invariant differential operators on Minkowski space and their curved analogues. *Commun. Math. Phys.* **109** (1987), 207–228.

[54] M. G. Eastwood and A. M. Pilato. On the density of twistor elementary states. Preprint.

[55] M. G. Eastwood. The twistor realization of discrete series. *Twistor Newsl.* **22** (1986), 39–40.

[56] M. G. Eastwood. On the weights of conformally invariant operators. *Twistor Newsl.* **24** (1987), 20–23.

[57] M. G. Eastwood. The Einstein bundle of a non-linear graviton. *Twistor Newsl.* **24** (1987), 3–4.

[58] M. G. Eastwood. The Penrose transform for curved ambitwistor space. *Quart. Jour. Math.* **39** (1988), 427–441.

[59] M. G. Eastwood and C. R. LeBrun. Fattening complex manifolds. Preprint.

[60] T. J. Enright and N. R. Wallach. Notes on homological algebra and representations of Lie algebras. *Duke Math. J.* **56** (1980), 1–14.

[61] T. Enright, R. Howe, and N. Wallach. A classification of unitary highest weight modules. In: *Representation Theory of Reductive Lie Groups*. (Proc. Utah Conf. 1982 (ed. P. C. Trombi) Prog. Math. vol. 40. Birkhäuser 1983) pp. 97–143.

[62] A. Ferber. Supertwistors and conformal supersymmetry. *Nucl. Phys.* **B132** (1978), 55–64.

[63] M. J. Field. *Several Complex Variables and Complex Manifolds*. Vols. I and II. L.M.S. Lecture Notes vols. 65 and 66. Cambridge University Press 1982.

[64] M. Fierz. Über den Drehimpuls von Teichen mit Ruhemasse null und bebliebigem Spin. *Helv. Phys. Acta* **13** (1940), 45–60.

[65] O. Foster and K. Knorr. Ein Beweis des Grauertschen Bilgarbenstaztes nach Ideen von B. Malgrange. *Man. Math.* **5** (1971), 19–44.

[66] H. Garland and J. Lepowsky. Lie algebra homology and the MacDonald–Kac formulae. *Inv. Math.* **34** (1976), 37–76.

[67] M. L. Ginsberg. A cohomological scalar product construction. In: *Advances in Twistor Theory* (eds. L. P. Hughston and R. S. Ward. Research Notes in Math. vol 37. Pitman 1979) pp. 293–300.

[68] M. L Ginsberg. *A Cohomological Approach to Scattering Theory*. D.Phil. thesis. Oxford University 1980.

[69] R. Godement. *Topologie Algébrique et Théorie des Faisceaux*. Hermann 1964.

[70] A. R. Gover. Conformally invariant operators of standard type. *Quart. J. Math.* 197–208 **40** (1989).

[71] H. Grauert. Ein Theorem der analytischen Garbentheorie und die Modulräume komplexen Strukturen. *Math. Publ. I.H.E.S.* **5** (1960).

[72] P. A. Green, J. Isenberg and P. B. Yasskin. Non self-dual gauge fields. *Phys. Lett.* **B78** (1978), 462–464.

[73] P. A. Griffiths and J. Adams. *Algebraic and Analytic Geometry*. Math. Notes vol. 13. Princeton University Press 1974.

[74] P. A. Griffiths and J. Harris. *Principles of Algebraic Geometry*. Wiley 1978.

[75] L. Gross. Norm invariance of mass-zero equations under the conformal group. *J. Math. Phys.* **5** (1963), 687–695.

[76] A. Grothendieck. *Local Cohomology* (notes by R. Hartshorne) Lect. Notes in Math. vol. 41. Springer 1967.

[77] V. Guillemin and S. Sternberg. *Symplectic Techniques in Physics*. Cambridge University Press 1984.

[78] R. C. Gunning and H. Rossi. *Analytic Functions of Several Complex Variables*. Prentice-Hall 1965. Wiley-Interscience 1970.

[79] Harish Chandra. Some applications of the universal enveloping algebra of a semi-simple Lie algebra. *Trans. A.M.S.* **70** (1951), 28–96.

[80] R. Hartshorne. *Algebraic Geometry*. Grad. Texts in Math. vol. 52. Springer 1977.

[81] F. R. Harvey. *Spinors and Calibrations*. Academic Press 1989.

[82] S. Helgason. *The Radon Transform*. Prog. Math. vol. 5. Birkhäuser 1980.

[83] S. Helgason. *Differential Geometry, Lie Groups and Symmetric Spaces*. Academic press 1978.

[84] G. M. Henkin and Yu. I. Manin. Twistor description of classical Yang–Mills–Dirac fields. *Phys. Lett.* **B95** (1980), 405–408.

[85] H. Hiller. *Geometry of Coxeter Groups*. Research Notes in Math. vol. 54. Pitman 1984.

[86] N. J. Hitchin. Linear field equations on self-dual spaces. *Proc. R. Soc. London* **A370** (1980), 173–191.

[87] A. P. Hodges and S. A. Huggett. Twistor diagrams. *Surv. High Energy Phys.* **1** (1980), 333–353.

[88] A. P. Hodges. Twistor diagrams. *Physica* **A114** (1982), 157–175.

[89] G. T. Horowitz. Introduction to string theories. In: *Topological Properties and Global Structure of Space-Time* (eds P. G. Bargmann and V. De Sabbata. NATO ASI series B. Physics; vol. 138. 1986) pp. 83–108.

[90] S. A. Huggett and K. P. Tod. *An Introduction to Twistor Theory*. L.M.S. Student Texts vol. 4. Cambridge University Press 1985.

[91] L. P. Hughston and R. S. Ward (eds). *Advances in Twistor Theory.* Research Notes in Math. vol 37. Pitman 1979.

[92] L. P. Hughston and T. R. Hurd. A $\mathbf{CP}^5$ calculus for space-time fields. *Phys. Rep.* **100** (1983), 273–326.

[93] L. P. Hughston and L. J. Mason (eds). *Further Advances in Twistor Theory.* Pitman, (1990) (to appear).

[94] J. E. Humphreys. *Introduction to Lie Algebras and Representation Theory.* Grad. Texts in Math. vol. 9. Springer 1972.

[95] J. E. Humphreys. *Linear Algebraic Groups.* Grad. Text. Math. vol. 21. Springer 1975.

[96] H. P. Jakobsen and M. Vergne. Wave and Dirac operators and representations of the conformal group. *J. Funct. Anal.* **24** (1977), 52–106.

[97] D. Kazhdan and G. Lusztig. Representations of Coxeter groups and Hecke algebras. *Inv. Math.* **53** (1979), 165–184.

[98] G. R. Kempf. The Grothendieck-Cousin complex of an induced representation. *Ad. Math.* **29** (1978), 301–396.

[99] R. Kiehl and J.-L. Verdier. Ein einfacher Beweis des Kohärenzsates von Grauert. *Math. Ann.* **195** (1971), 24–50.

[100] S. Kobayashi and T. Nagano. On filtered Lie algebras and geometric structures. *J. Math. Mech.* **13** (1964), 875–907.

[101] K. Kodaira. *Complex Manifolds and Deformation of Complex Structures.* Grundlehren Band 283. Springer 1986.

[102] B. Kostant. Lie algebra cohomology and the generalized Borel-Weil theorem. *Ann. Math.* **74** (1961), 329–387.

[103] C. Kozameh, E. T. Newman, and K. P. Tod. Conformal Einstein spaces. *G. R. G.* **17** (1985), 343–352.

[104] H. B. Laufer. On Serre duality and envelopes of holomorphy. *Trans. A.M.S.* **128** (1967), 414–436.

[105] C. R. LeBrun. Spaces of complex null geodesics in complex-Riemannian geometry. *Trans. A.M.S.* **278** (1983), 209–231.

[106] C. R. LeBrun. Ambi-twistors and Einstein's equations. *Classical Quantum Grav.* **2** (1985), 555–563.

[107] C. R. Le Brun. Thickenings and conformal gravity. Preprint, 1989.

[108] J. Leiterer. The Penrose transform for bundles non-trivial on the general line. *Math. Nachr.* **112** (1983), 35–67.

[109] J. Leiterer. Subsheaves in bundles on $\mathbf{P}_n$ and the Penrose transform. Preprint.

[110] J. Lepowsky. A generalization of the Bernstein–Gelfand–Gelfand resolution. *J. Algebra* **49** (1977), 496–511.

[111] D. B. Lichtenberg. *Unitary Symmetry and Elementary Particles.* Academic Press 1970.

[112] I. G. MacDonald. *Symmetric Functions and Hall Polynomials.* Oxford University Press 1979.

[113] Yu. I. Manin. Gauge fields and holomorphic geometry. *Curr. Probl. Math. (Akad. Nauk S.S.S.R.)* **17** (1981), 3–55 (in Russian) and *J. Sov. Math.* **21** (1983), 465–507 (in English).

[114] J. A. McLennan, Jr.. Conformal invariance and conservation laws for relativistic wave equations for zero mass. *Nuovo Cim.* **3** (1956), 1360–1379.

[115] T. Ochiai. Geometry associated with semisimple flat homogeneous space. *Trans. A.M.S.* **152** (1970), 159–193.

[116] R. Penrose. Twistor algebra. *J. Math. Phys.* **8** (1967), 345–366.

[117] R. Penrose. Structure of space-time. In: *Battelle Rencontres: 1967* (eds. C. M. DeWitt and J. A. Wheeler. Benjamin 1968) pp. 121–235.

[118] R. Penrose. Twistor quantization and curved space-time. *Int. J. Theor. Phys.* **1** (1968), 61–99.

[119] R. Penrose. Solutions of the zero-rest-mass equations. *J. Math. Phys.* **10** (1969), 38–39.

[120] R. Penrose and M. A. H. MacCallum. Twistor theory: an approach to the quantisation of fields and space-time. *Phys. Rep.* **6** (1972), 241–316.

[121] R. Penrose. Twistor theory, its aims and achievements. In: *Quantum Gravity: an Oxford Symposium* (eds C. J. Isham, R. Penrose, and D. W. Sciama. Clarendon Press 1975) pp. 268–407.

[122] R. Penrose. Twistors and particles: an outline. In: *Quantum Theory and the Structure of Space-time* (eds L. Castell, M. Drieschner, and C. F. von Weizsäcker. Carl Hanser 1975) pp. 129–145.

[123] R. Penrose. Non-linear gravitons and curved twistor theory. *G. R. G.* **7** (1976), 31–52.

[124] R. Penrose. On the twistor description of massless fields. In: *Complex Manifold Techniques in Theoretical Physics* (eds D. E. Lerner and P. D. Sommers. Research Notes in Math. vol. 32. Pitman 1979) pp. 55–91.

[125] R. Penrose. Twistor functions and sheaf cohomology. In: *Advances in Twistor Theory* (eds L. P. Hughston and R. S. Ward. Research Notes in Math. vol. 37. Pitman 1979) pp. 25–36.

[126] R. Penrose and R. S. Ward. Twistors for flat and curved space-time. In: *General Relativity and Gravitation*, vol. II (ed A. Held. Plenum 1980) pp. 283–328.

[127] R. Penrose and W. Rindler. *Spinors and Space-time*, vol. I: Two-spinor calculus and relativistic fields. Cambridge University Press 1984.

[128] R. Penrose and W. Rindler. *Spinors and Space-time*, vol. II: Spinor and twistor methods in space-time geometry. Cambridge University Press 1986.

[129] R. Pool. *Yang–Mills Fields and Extension Theory. A.M.S. Mem.* **358** (1987).

[130] M. Roček. An introduction to superspace and supergravity. In: *Superspace and Supergravity* (eds S. W. Hawking and M. Roček. Cambridge University Press 1981) pp. 71–131.

[131] A. Rocha-Caridi. Splitting criteria for **g**-modules induced from a parabolic and the Bernstein–Gelfand–Gelfand resolution of a finite dimensional irreducible **g**-module. *Trans. A.M.S.* **262** (1980), 335–366.

[132] P. Schapira. *Microdifferential Systems in the Complex Domain.* Grundlehren Band 269. Springer 1985.

[133] W. Schmid. *Homogeneous complex manifolds and representations of semisimple Lie groups.* Ph.D. thesis. University of California, Berkeley 1967.

[134] W. Schmid. Homogeneous complex manifolds and representations of semisimple Lie groups. *Proc. Nat. Acad. Sci. U.S.A.* **59** (1968), 56–59.

[135] W. Schmid. Representations of semisimple Lie groups. In: *Proceedings of the International Congress of Mathematicians: Helsinki 1978* (ed. O. Lehto) pp. 195–208.

[136] W. Schmid. On a conjecture of Langlands. *Ann. Math.* **93** (1971), 1–42.

[137] W. Schmid. L^2-cohomology and the discrete series. *Ann. Math.* **103** (1976), 375–394.

[138] W. Schmid. Representations of semi-simple Lie groups. In: *Representation theory of Lie Groups* (ed. M. F. Atiyah. L.M.S. Lecture Notes vol. 34. Cambridge University Press 1979) pp. 185–235.

[139] C. S. Seshadri. Quotient spaces modulo reductive algebraic groups. *Ann. Math.* **95** (1972), 511–556.

[140] J.-P. Serre. Un théorème de dualité. *Commun. Math. Helv.* **29** (1955), 9–26.

[141] M. A. Singer. Duality in twistor theory without Minkowski space. *Math. Proc. Camb. Philos. Soc.* **98** (1985), 591–600.

[142] M. A. Singer. The generalized Penrose transform—global aspects. Preprint.

[143] M. A. Singer. A duality theorem for the de Rham theory of a family of manifolds. Preprint.

[144] G. Sorani and V. Villani. q-complete spaces and cohomology. *Trans. A.M.S.* **125** (1966), 432–448.

[145] D. C. Spencer. Overdetermined systems of linear partial differential equations. *Bull. A.M.S.* **75** (1969), 179–239.

[146] R. F. Streater and A. S. Wightman. *PCT, Spin and Statistics, and all that.* Benjamin 1964.

[147] M. E. Taylor. *Noncommutative Harmonic Analysis.* Mathematical Surveys and Monographs vol. 22. A.M.S. 1986.

[148] A. Trautman. Yang-Mills theory and gravitation: a comparison. In: *Geometric Techniques in Gauge Theories* (eds. R. Martini and E. M. de Jager. Lecture Notes in Math. vol. 926. Springer 1982) pp. 179–189.

[149] V. S. Varadarajan *Lie Groups, Lie Algebras, and their Representations.* Prentice-Hall 1974.

[150] Verma D.-N. Structure of certain induced representations of complex semisimple Lie algebras. *Bull. A.M.S.* **74** (1968), 160–166.

[151] D. A. Vogan. *Representations of Real Reductive Lie Groups.* Prog. Math. vol. 15. Birkhäuser 1981.

[152] H. C. Wang. Closed manifolds with complex homogeneous structure. *Am. J. Math.* **76** (1954), 1–32.

[153] R. S. Ward. On self-dual gauge fields. *Phys. Lett.* **A61** (1977), 81–82.

[154] R. S. Ward. Self-dual space-times with cosmological constant. *Commun. Math. Phys.* **78** (1980), 1–17.

[155] R. S. Ward and R. O. Wells, Jr.. *Twistor Geometry and Field Theory.* Cambridge University Press 1989.

[156] G. Warner. *Harmonic Analysis on Semi-simple Lie Groups I and II.* Grundlehren Bände 188 und 189. Springer 1972.

[157] R. O. Wells, Jr. *Differential Analysis on Complex Manifolds.* Grad. Texts in Math. vol. 65. Springer 1980.

[158] R. O. Wells, Jr. Complex manifolds and mathematical physics. *Bull. A.M.S.* **1** (1979), 296–336.

[159] B. G. Wybourne. *Symmetry Principles and Atomic Spectroscopy.* Wiley-Interscience 1970.

[160] J. Wolf. The action of a real semi-simple group on a complex flag manifold I: Orbit structure and holomorphic arc components. *Bull. A.M.S.* **75** (1969), 1121–1237.

[161] E. Witten. An interpretation of classical Yang–Mills theory. *Phys. Lett.* **B77** (1978), 394–398

[162] E. Witten. Twistor-like transform in ten dimensions. *Nucl. Phys.* **B266** (1986), 245–264.

[163] N. M. J. Woodhouse. *Geometric Quantization.* Oxford University Press 1980.

[164] G. Zuckerman. Tensor products of finite and infinite dimensional representations of semisimple Lie groups. *Ann. Math.* **106** (1977), 295–308.

INDEX